Pharmacology and Abuse of Cocaine, Amphetamines, Ecstasy and Related Designer Drugs

Enno Freye

Pharmacology and Abuse of Cocaine, Amphetamines, Ecstasy and Related Designer Drugs

A Comprehensive Review on their Mode of Action, Treatment of Abuse and Intoxication

In Collaboration with Joseph V. Levy[†]

WITHDRAWN

Enno Freye, MD, PhD
Heinrich-Heine-University Düsseldorf
Germany
enno.freye@uni-duesseldorf.de

In collaboration with
Joseph V. Levy[†], PhD
University of the Pacific
School of Dentistry
2155 Webster St.
San Francisco CA 94115
USA

ISBN 978-90-481-2447-3   e-ISBN 978-90-481-2448-0
DOI 10.1007/978-90-481-2448-0
Springer Dordrecht Heidelberg London New York

Library of Congress Control Number: 2009926516

© Springer Science+Business Media B.V. 2009
No part of this work may be reproduced, stored in a retrieval system, or transmitted in any form or by any means, electronic, mechanical, photocopying, microfilming, recording or otherwise, without written permission from the Publisher, with the exception of any material supplied specifically for the purpose of being entered and executed on a computer system, for exclusive use by the purchaser of the work.

Printed on acid-free paper

Springer is part of Springer Science+Business Media (www.springer.com)

Contents

Introduction	1
Incidence of Illicit Use of Cocaine and Other Drugs	1
Part I How It All Started with Cocaine	5
Botanical Characteristics of Coca Leaves	7
Production of the Coca Leaf	9
Cocaine: History of Use	13
Coca Consumption with Wine	15
Coca in Refreshing Beverages	15
Regulatories for Use of Cocaine	16
Popularity of Cocaine as a Drug for Medical Treatment	19
Use of Cocaine as a Local Anesthetic	21
The Different Forms of Cocaine	25
The Different Types of Alkaloids in Coca	27
The Making of Cocaine in the Jungle	29
Steps Refining Coca Paste to Cocaine Base	31
Trafficking of Cocaine: The Legal Approach	33
Profit Making and Trafficking of Cocaine: The Illegal Approach	33
Smuggling of Cocaine	35
Diluting or "Cutting" of Cocaine	36
Ascertaining the Purity of Cocaine and the Cutting Agent	37
Economics of Cocaine	40
Freebase Cocaine: High Bioavailability with Increase in Potency	43
Freebasing of Cocaine: The Ether Wash Method	44
Freebasing of Cocaine: The Baking Soda Method	47

Pharmacology of Cocaine	49
Pharmaceutical Cocaine	49
Physicochemical Properties of Cocaine	50
Decomposition of Cocaine	51
Cocaine, as a Local Anesthetic: Mechanism of Action	51
Cocaine's CNS Effects: Mechanism of Action	54
Emotional Effects of Cocaine	58
Addictive Properties of Cocaine: Mode of Action	61
Psychological Dependence	63
Cocaine Intoxication: Strategy for Treatment	65
Phase of Early Stimulation	66
Phase of Advanced Stimulation	66
The Depressive Phase	66
Basics in Treatment of Cocaine Overdose (OD)	66
Advanced Treatment of Cocaine OD	67
Selective Approach to Cocaine OD	67
Avoiding Procedures in Cocaine OD	68
Special Pathologies in Chronic Cocaine Use	69
The Toxic Cocaine Delirium	69
Cocaine and Myocardial Infarction	70
Treatment of Cocaine-Related Myocardial Ischemia (MI)	71
Secondary Prevention of Cocaine-Related MI	72
Cocaine-Related Rhabdomyolysis	73
Cocaine and Cerebrovascular Ischemia/Hemorrhage	74
Summary of Advanced Treatment in Cocaine OD	75
Treatment of Tachyarrhythmia	75
Treatment of Ventricular fibrillation	75
Treatment of Hypertension	76
Treatment of Hyperthermia	76
Treatment of Seizures	76
Lethal Doses of Cocaine	77
Chronic Toxicity of Cocaine use	79
Summary of Acute-Chronic Effects of Cocaine	83
Duration of Withdrawal Symptoms Following Cocaine Abuse	85
Cocaine Use in Pregnancy	87
Treatment Options in Cocaine Abuse	88
Detoxification	88
Cocaine-Free Environment	89

Hospitalization or Outpatient Therapy	89
Pharmacological Therapy	89
Use of Antidepressants to Relieve Post-Cocaine Depression	92
Non-pharmacological Treatmenet of Cocaine Addiction	96

New Options in the Treatment of Cocaine Dependency 99
The Antiepileptic Gamma Vinyl-GABA (GVG) 99
Disulfiram (Antabuse®) 100
Use of an Anti-cocaine Vaccine 102
Vaccination of the Cocaine User 102
Newer Dopamine Reuptake Inhibitors 103
    GBR-12909 (Vanoxerine) 103
    Venlafaxine (Effexor®) and Ecopipam 103
Outlook Regarding Therapy in Cocaine Abuse 104

References 105

Part II   Methamphetamine-an Old Aquaintance in New Clothes 109
History of Methamphetamine 109
From Amphetamine to Methamphetamine 110

Mode of Action of Methamphetamine 115
Interaction of Methamphetamine with other Stimulants 117

Pharmacology of Methamphetamine 119
Prescribing Methamphetamine 122
Chemistry of Methamphetamine 123

Crystal Methamphetamine 125
Addictive Properties of Methamphetamine 125

**Treatment Options in Methamphetamine Addiction
and Withdrawal** 131
Medical Treatment in Methamphetamine Related Depression 132
Non-Medical Treatment in Methamphetamine Related Depression 132
Treatment Preventing Methamphetamine Relapse 132

Amphetamine Derivatives as Appetite Suppressants 135

References 137

**Part III   MDMA (Ecstasy): from Psychotherapy
               to a Street Drug of Abuse** 139
History of MDMA 139

Incidence of Illicit Use of Ecstasy 143
Medical Use of Ecstasy in Assisting Psychotherapy 143

Pharmacology of Ecstasy (MDMA) 147

Pharmacological Effects of MDMA in Man 151
Onset of Action and Duration of Action 151
The Psychological Effects of MDMA 151
    Entactogenesis ("Touching Within") 152
    Empathogenesis 152
    Enhancement of Senses 152
Abuse Liability and Psychomotor Performance 153
The Physical Effects of MDMA 153
    Cardiovascular Effects of MDMA 154
    Neuroendocrine Effects of MDMA 154
    Ocular Effects of MDMA 155
    Lethal Effects of MDMA 155
    Effects in Chronic Use of MDMA 155
Chronic MDMA Use and Neurotoxicity 156

Pharmacokinetics of MDMA 161
Primary Metabolism of MDMA 161
Secondary Metabolism of MDMA 163
Interactions and Overdose with MDMA 164
Acute Intoxication with MDMA 165
    Hyperthermia 166
    Cardiovascular Effects 166
    Cerebrovascular Effects 167
    Neuroendocrine Effects 167
    Hepatotoxicity 168
    Psychopathology 168
Treatment of MDMA-Related Toxicity 168
    Hyperthermia 169
    Cardiovascular Treatment 170
    Cerebrovascular Treatment 170
    Neurologic Treatment 171

Appendix   Analogues of MDMA and Related Compounds 173
MDA (3,4-methylenedioxyamphetamine) 173
MDE or MDEA (N-ethyl-methylenedioxyamphetamine) 173
MMDA (3-methoxy-4,5-methylenedioxyamphetamine) 173
2-Methoxy-3,4-methylenedioxyphenethylamine 174
MBDB (N-methyl-1-(1,3-benzodioxol-5-yl)-N-methylbutan-2-amine) 174

References 175

Contents

Part IV Designer Drugs and their Abuse 181

History of Designer Drugs 183
Well-Known Designer Drugs of the Past Years 186

Gamma-Hydroxybutyric (GHB) Acid 191
History of GHB 191
Medical Use of γ-Hydroxybutyrate 192
Pharmacokinetics of GHB 194

The Pharmacology of γ-Hydroxybutyric Acid (GHB) 195
Mode of Action of GHB 195
    Dose-Related CNS Effects of GHB 197
    Cardiovascular Effects of GHB 197
    Respiratory Effects of GHB 198
    Neuroendocrine Effects of GHB 198
    Sleep Pattern Following GHB 198

Use of GHB in Clinical Medicine 199
GHB for Sedation and Anesthesia 199
GHB for Cerebral Protection 199
GHB in Narcolepsy and Insomnia 200
GHB in Alcohol and Opiate Withdrawal 200
Metabolism of γ-Hydroxybutyrate(GHB) 201

Abuse Potential and Intoxication with GHB 203
Incidence of Adverse Effects in GHB Intoxication 203
Principles in GHB Intoxication 204
Effects of Acute GHB Intoxication 205
    Cardiovascular Effects with GHB Intoxication 205
    Respiratory Effects with GHB Intoxication 205
    Psychotic Effects with GHB Intoxication 206
    Ocular Effects with GHB Intoxication 206
    Metabolic Effects with GHB Intoxication 206
    Gastrointestinal Effects with GHB Intoxication 206
    Body Temperature in GHB Intoxication 206
    Muscular Disorders in GHB Intoxication 207
    Miscellaneous Effects in GHB Intoxication 207
Withdrawal and Tolerance Following Use of GHB 207
Treatment Following GHB Intoxication 208
    First-Line Treatment in GHB Overdose 208
    Cardiovascular Withdrawal Treatment Following GHB Overdose 209

Benzylpiperazine (BZP) as a Designer Drug 211
Pharmacology of Benzylpiperazine (BZP) 211
Addictive Effects of Benzylpipcrazine 212

Natural Products with Abuse Potential	215
Datura a Hallucinogenic	215
Toxicity of Datura Stramonium	217
Overdose of Datura	217
Dimethyltryptamine (DMT) a Psychedelic	219
The Mushroom Psilocybin with Psychedelic Properties	221
Toxicity of Psilocybine	221
Pharmacological Effects of Psilocybine	221
Ibogaine, Psychedelic Molecule with Anti-addictive Properties	225
Pharmacology of Ibogaine	226
Peyote, a Mescaline-Containing Cactus	227
Behavioral Effects Produced by Mescaline	227
LSD, a Semisynthetic Psychedelic Drug	229
Pharmacology of LSD	229
Research with LSD	230
The Legal Status of LSD	231
5-MeO-DIPT, a Psychedelic Structurally Related to Psiloc(yb)in	233
Pharmacology of 5-MeO-DIPT	233
Legal Status of 5-MeO-DIPT	234
References	235
Part V How to Demask the Patient with an Aberrant Drug Use	243
Stigma of Persons with a Potential Drug Addictive Behavior	243
The Alcohol Addictive Patient	244
Signs of Alcohol Co-abuse in Addiction	244
Positive Signs and Symptoms of Hard Drug Abuse	247
Signs and Symptoms for Abusive Drug Behavior	247
Urine Drug Screening (UDS): Identifying the Person with Illicit Use	251
Prerequisites for Use of UDS	251
Types of Urine Drug Testing	253
Half-Life of Detection and Cut-Offs in Urine Drug Screening	255
Interpretation of Positive Urine Drug Tests (UDT)	256
Considerations When Having a Positive Urine Drug Test	256
Considerations When Having a Negative Urine Drug Test	260

Contents

Type of Tests for Urine Drug Testing (UDT) 261
The Dip Test 261
The Drop Test 261
    How to Interpret the Results 261
Identification of the Methadone Metabolite in Substitution Therapy 262
Single Test Strips 265
Test-Cup for Permanent Storage 266
Detecting Manipulations of Urine Samples in UDT 267

Analysis of Saliva, Hair and Sweat for Drug Testing 269
Hair Analysis for the Detection of Abused Drugs and Medications 269
Analysis of Sweat for Abused Drugs and Medications 271
    SmartClip® Test for Saliva Testing 272
    OraLab® Test for Saliva Testing 274
    OraTube® for Fluid Testing 274
    DRUGWIPE® 5+ for Drug Testing of Saliva 275
    Dräger Drug Test 5000 278
    Trace Wipe Test for Identification of Cocaine Smuggling or Abuse 279
    Professional Equipment for Drug Analysis of Liquids/
    Powder/Tablet 279
    Drug Testing in Vapor 282

References 283

Appendix   Common Street Names of Cocaine and Methamphetamine used as a Jargon 287

Index 297

Introduction

There are about 1.6 million new drug abusers per annum in Europe, suggesting that there is an impact need in knowledge when patients are being worked up (Fig. 1). The following book is intended to give the practitioner decisions at hand to choose the most effective treatment, consider potential drug abuse in certain pathological conditions and reflect on the possibility of interaction of abused drugs with their regular medications.

An Incan prophesy says: *"If the white man takes coca, one day a white powder will destroy him"*. As be seen in the next decades to come, the proliferation of "white powders" in the society, especially cocaine and amphetamine, will reach epidemic proportions. The effect of these substances in terms of social and economic costs is enormous. It is believed by some sociologists that drug abuse is the single greatest danger faced by the Western World. While perhaps an overstatement, drug abuse is a growing menace, fueled in part by greed, social attitudes, and a long history of involvement with psychoactive substances.

The following chapters will cover these "white powders" known as cocaine, amphetamines or ecstasy. Its origins will be traced, from plant to drug, its pharmacology explored and its impact on the human or the animal described with respect to abuse, overdose and dependence.

Incidence of Illicit Use of Cocaine and Other Drugs

According to the 2006 National Survey on Drug Use and Health (NSDUH), approximately 35.3 million Americans aged 12 and older had tried cocaine at least once in their lifetimes, representing 14.3% of the population aged 12 and older. Approximately 6.1 million (2.5%) has used cocaine in the past year and 2.4 million (1.0%) had used cocaine within the past month. Data from the 2006 NSDUH also indicate that there were 977,000 persons aged 12 or older who had used cocaine for the first time within the past 12 months, averaging to approximately 2,700 initiates per day. This estimate was not significantly different from the number in 2005 (872,000). Among students surveyed as part of the 2007 Monitoring the Future

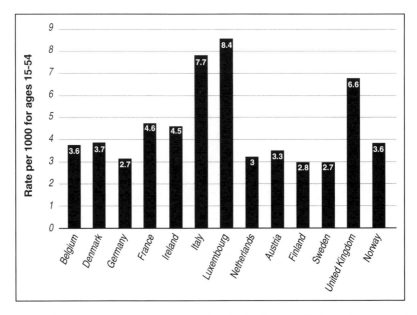

Fig. 1 Annual report on the state of drugs problem in the European Union Adapted from the European Monitoring Centre for Drugs and Drug Addiction (EMCDDA) 2001

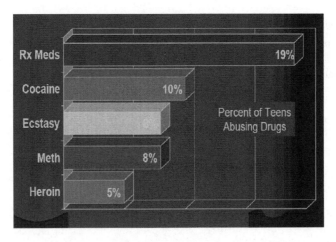

Fig. 2 Percent of teens reporting cocaine use from 2006–2007 in combination or alone cover at least 10%. Source: Partnership Attitude and Tracking Study (PATS) 2005

study, 3.1% of 8th graders, 5.3% of 10th graders, and 7.8% of 12th graders reported lifetime use of cocaine. In 2006, these percentages were 3.4%, 4.8%, and 8.5%, respectively (Fig. 2 and Table 1).

Table 1 Teenage Rx drug use and polydrug abuse trends. Source: National Center on Addiction and Substance Abuse

Teens who abuse Rx drugs are:
- 2 times more likely to abuse alcohol
- 5 times more likely to abuse marijuana
- 12 times more likely to abuse heroin
- 15 times more likely to abuse ecstasy
- 21 times more likely to abuse cocaine

Common used names for disguising illicit drug consumption

1. The bowls and baggies of Rx Drugs are often called *"Trail Mix"*.
2. Some websites and chat rooms refer to pills by color rather than brand, name, content, or potency.
3. *"Pharming"* refers to use/abuse of Rx meds to get high.
4. *"Pharming Parties"* refer to parties wherein Rx drugs are exchanged among and between youth.

Part I
How It All Started with Cocaine

Before there was cocaine, in the beginning there was coca, *"The divine plant of the Incas"*. The section will describe the plant, what it looks like, where it grows, what is in it (i.e. the alkaloids), its history, who uses it and how it is abused.

Coca must be considered separately from its principal alkaloid cocaine, because there's much more to coca than just cocaine. The coca plant, being a rather innocuous looking bush (Fig. 3), is the source of cocaine and several other alkaloids. It is believed that the name originates from the Aymara Indians of pre-Inca Bolivia (prior to the tenth century AD), where coca literally means *"plant"* or *"tree"*, suggesting that it entails that "the plant" is even "the plant of all plants".

The only identified coca plants that have attributed to the production of cocaine are the Erythroxylum coca and the Erthroxylum novogranatense plants. In this regard coca is a member of the genus Erythroxylum, in which are found at least 17 species that are known to contain the alkaloid cocaine. The two species that are widely cultivated for their cocaine and other alkaloids are:

1. **Erythroxylum Coca** (E. coca); the most prized because its cocaine content is the highest (up to 90% of the total alkaloidal content or 1–1.8% of the total weight of the leaf). E. coca is also known as the Huanaco or Bolivian leaf. A variety of E. coca (Ipadu) grows in the Amazon region, but it has a much lower cocaine content.
2. The other principal species of Erythroxylum is **Erythroxylum Novogranatense**. This plant grows mostly in the mountains of Peru, Ecuador and Colombia. There are several varieties of this species, two of which are:

    (a) **Truxillense** (Trujillo or Peruvian leaf) which is the variety used to flavor beverages such as Coca Cola, and the
    (b) **Javanese** variety (Java Coca) from Indonesia, which supplied most of the world's pharmaceutical cocaine prior to World War II. Illicit growers use whatever leaves they can get, and often transplant varieties of coca plants to areas previously not used for cultivation. The Amazon basin is one such area.

Fig. 3 The coca plant has elliptical leaves commonly with a small white flowering bud

Erythroxylum novogranatense is mostly grown in the Colombian and Central America countries. In Columbia it grows in the Sierra Nevada de Santa Marta Mountains where there is extreme heat and drought. It differs from *Erythroxylum coca* in its color, texture, and odor. Its cocaine content is much lower then the Erythoxylum coca plant. Although the plants may have different appearance, the cocaine alkaloid in the two coca plants set them apart from others. *Erythroxylum coca*, widely known as Huanco, grows along the eastern slopes of Bolivia and Peru. It grows in altitudes from 500–1,500 m. It spreads a hay-like odor and only grows in extremely humid and wet conditions. Soil composition and weather conditions determine the strength of the coca leaves as well. The cocaine production comes in 95% from the *Erythroxylum coca* (Fig. 4).

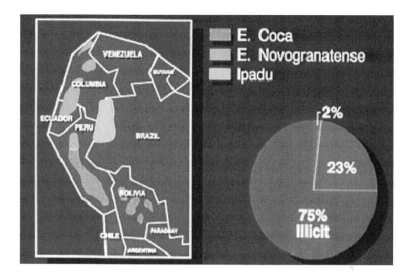

Fig. 4 The different countries in South America involved in the production of illicit cocaine

Botanical Characteristics of Coca Leaves

Coca is primarily a mountain plant, although it has been grown in a variety of climatic zones. These plants live a long time, and are hard to destroy. The bushes are normally cut back to 3–6 ft in order to facilitate picking. There are only a few plants per acre, but a lot of leaves are needed to produce a small amount of cocaine. Attempts to grow E. coca in other soil and different climate conditions failed to produce good yields. Hence, farming in the US and/or Europe was given up. This is unlike some other illegal plants such as marijuana and certain mushrooms, which, because of their yields, have been grown in commercial quantities.

Coca plants are usually found on terraced hillsides where they have been cultivated for centuries, and so little grows wild anymore. A farm for growing coca is called a *"cocale"*, while growers and chewers of coca are known as *"coqueros"* (Fig. 5).

Fig. 5 Terrace-like neat rows of coca plants are carefully tended. This is rarely the case in illicit plots found in forest areas or in the jungle. A good "cocale" may have as many as 7,000 plants per acre

Production of the Coca Leaf

The map in Fig. 6 shows the main countries in Latin America involved in the production of coca leaves. Based on metric tons of coca leaves harvested annually, and although exact figures are hard to verify, the following may give some idea as to the analysis, all of which accounts to a total of 112,000 t (Fig. 7).

There are three harvest times a year in March, June, October where 4 oz of leaves per bush can be obtained during the main harvest. It takes up to 2 days for two people to pick 12 kg and the grower (coquero) roughly makes $2.20/kg.

Also, illicit growers may not get as good a yield from their plants due to their use of less potent species, and growing the plants under less favorable climatic conditions such as the Amazon basin. As can be seen, the peasant growers (coqueros) make little from their harvest, and they must sell their leaves to the government.

There is early archaeological evidence (Ecuador) for the use of Coca (Divine Plant of the Incas), but the history of coca precedes the Incas by thousands of years. Mummies with coca leaves wrapped up with them found in Ecuador and Peru confirm that coca was cultivated as far back as 3000 BC. Each successive tribe had their own legends as to the source of coca, but it was the Incas who raised coca to new heights of religious significance after the thirteenth century AD. The Inca state controlled the cultivation and distribution of coca, reserving it for the upper classes, and the workers, presumably to get more work out of them; "Deaden man's hunger and give him strength to live" (Fig. 8).

Today, coca is widely still used in the mountains of what once was the Inca empire. The government primarily controls coca, since coca leaves must be sent by growers (coqueros) to a government-owned factory. There the leaves are being pressed after which they are stored in a government's owned coca warehouse.

Coca leaves have been chewed by South American Indians for many thousands of years to induce a mild, long-lasting euphoria. The Incas respected coca as it was used in magical ceremonies and initiation rites. In the Inca period, the sacred leaf was regarded as far too good for ordinary Indians. The invading Spanish conquistadors were more practical. They were impressed of coca's efficacy as a stimulant and believed that the herb was so nutritious and invigorating that the Indians labored whole days without anything else. The Spanish also needed native labor in their silver mines. Work in the mines was extremely arduous, and taking coca reduced

Fig. 6 Main countries involved in coca leaf cultivation of Latin America

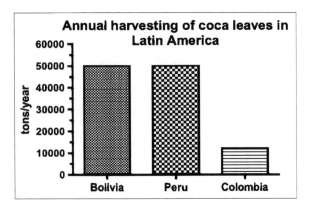

Fig. 7 The annual amount of coca leaf production in Latin America

Fig. 8 Indios chewing the coca leaf as a social drug on special occasions

appetite and increases physical stamina. Therefore, there was a great surge in coca use and the number of coqueros (coca-chewers). The chewing of coca is a well-defined practice and has been around for centuries, where the coca leaf has long been used as a stimulant by indigenous peoples of the Andean region. Coca was chewed in its nature leaf like state. The chewing of coca leaves had been reserved for the noble class and was usually associated with religious activities. Thus, the coca leaf soon began to be chewed by more then just the noble class. Indians began to chew the leaves to conquer hunger, improve muscle stamina, and to counter motion sickness and oxygen deprivation. The leaves aren't actually chewed, but rather moistened with saliva, softened by the addition of an alkali, such as lime (calcium carbonate – from vegetable ash or powdered sea shells) and sucked for about 30 min.

Coca leaf chewing today is more widespread than many appreciate. There are at least three million daily users of coca leaf in South America, primarily in Bolivia, Peru, Ecuador and Colombia consuming 2 oz (equivalent to 1/4 to 1/2 g cocaine) per day, which runs to over 25,000 t of coca leaves/year with an average daily cost of 13 cents/oz. Local laws vary and in some countries there is an official attitude of indifference to the remote Indian population using coca, even when it is officially banned elsewhere. In Bolivia and Peru the government "Coca Monopoly" controls the cultivation and distribution of coca. Both men and women use coca, however, in the mountains of Colombia women rarely and children never use it. This is also true in the Amazon basin where coca leaves are first toasted and then grounded into a powder before being put in the cheek together with lime. While the average daily consumption of coca leaves is about 2 oz, this translates roughly to a quarter to half-a-gram of cocaine per day. It must be emphasized that while this seems high even by the standards of cocaine abusers in the US, this amount of alkaloidal cocaine is taken slowly, over the course of several hours, and is hardly enough to sustain even mild plasma levels of cocaine in the user. As will be seen later, this is far from using the purified alkaloid cocaine by itself. In 1950, the United Nations began a campaign against the coca plant. In 1961 the UN Single Convention on Narcotic Drugs called for the abolition of coca chewing and the destruction of plants within 25 years (signed by Peru – approved by Bolivia). In 1986, however, the coca harvest was as big as it had ever been.

For legal chewing 25,000 t and for pharmaceutical purposes as well as a flavor in beverages 2,000 t are being used, which leaves 85,000 t available for illicit exploit. As can readily be seen, while the cultivation of coca for chewing by the local population, beverage flavoring and pharmaceutical manufacture is legal, the majority (75%) is available for diversion to the illicit manufacturers of cocaine. It is unfortunate that eradication efforts have been more successful in removing the legal plants and less so when in comes to the illegal ones. One side effect of this has been the animosity on the part of legal chewers, who must now pay more for their leaves or go without entirely, with the resultant disruption of a centuries-old way of life.

In addition to coca chewing, "Coca de mate" a tea made from the leaves is also popular, since drinking coca tea tends to soothe the stomach and is good for digestive problems.

Cocaine: History of Use

Coca has been used for ages as a food substitute, a stimulant, a medicine, as an aphrodisiac, a means to stay warm, and as a measure of distance running. An important factor in the spread of coca-chewing among Indians was the need for a food substitute when the Incan agricultural economy broke down due to inter tribal wars. Nutritional analysis shows that 100 g of coca leaves contain 305 cal, 18.9 g of protein, and 46.2 g of carbohydrates, and satisfies the recommended dietary allowances for calcium, iron, phosphorus, and vitamins A, B, C, and E. As a medicinal herb, coca has been used in treating a variety of ailments and diseases, being applied by shaman or medicine men in rites and ceremonies. Studies show that coca leaves have peripheral vasoconstrictive effects that reduces the amount of heat loss through the extremities and produces a higher central body temperature keeping the user warmer.

Coca leaves must be dried after harvesting or else they will lose their potency. Properly dried leaves will retain their potency for several years. Along with the discovery of tobacco and other crops unique to the New World, the Europeans were also introduced to coca. Beginning with the early explorers, who noticed the Indians in the Caribbean chewing coca leaves, and continuing with the conquest of the Incas by Pizarro (1531–1536), the white men were both fascinated and repulsed by this practice. At first, the use of coca was discouraged as evil, but in time, the Spanish learned that the Indians worked harder at high altitudes when they had coca, a fact that became especially obvious when a large silver mine was discovered in 1545 – at an altitude of 16,000 ft (Fig. 9).

By 1565, the King of Spain had issued an official approval for coca's use by the Indian laborers, and a heavy tax imposed on coca helped to build many churches and other buildings in what is now Bolivia and Peru. Accompanying the conquistadors were some men of science, doctors one of whom was Nicholas Monardes, who wrote the first scientific account of coca in about 1569. Interest in coca, however, did not reach similar heights as achieved by tobacco due in part to its inability to be smoked (Europeans didn't take to chewing). Other reasons coca didn't become as popular in Europe as did the tobacco plant, is that the leaves were improperly stored on the voyage from South America and that the leaves had lost most of their potency (Fig. 10).

Fig. 9 With the Spanish Conquest in 1507 Amerigo Vespucci wrote of coca chewing by Indians. In 1531 Pizarro arrives in Peru and conquers Incas territory. In 1545 silver mines were exploited with Indian laborers working at an altitude of 16,000 ft. In 1585 Phillip II of Spain approved use of coca for Indian laborers, and in 1569 the first scientific account of coca was written by Nicolas Monardes

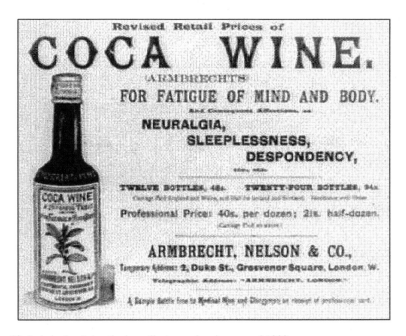

Fig. 10 Label advertising the ingredient coca in wine around 1900

Coca Consumption with Wine

The real soaraway success in Europe was "Vin Mariani". Launched in 1863, it was an extremely palatable coca wine developed by the Corsican chemist and entrepreneur, Angelo Mariani (1838–1914). Prior to this "Vin Mariani" Coca had remained largely obscure until it had been introduced as a product containing extract of coca leaves (not purified cocaine) and wine (2 oz of leaves to a pint of Bordeaux wine) in 1863. Mariani had first tried this new tonic on a depressed actress with spectacular results. Writing a book eulogizing coca, he gathered artifacts of, and material on, the coca-loving Incas. At home, he collected coke-taking paraphernalia. He also took up amateur horticulture and cultivated the coca plant in his garden. When "Vin Mariani" became the most popular of a host of coca-extract products being mass-marketed by Mariani, he became one of the greatest salesmen of his time. By sending "Vin Mariani" for free to famous personalities around the world, he solicited their "opinions". He then asked for photographs of them with their endorsements, which were published in various newspapers and magazines (Fig. 11).

Coca in Refreshing Beverages

One imitator of Mariani was the Atlanta pharmacist, John Styth Pemberton (Fig. 12). In 1885 he introduced "French Wine Coca" as an "Ideal Nerve and Tonic Stimulant". The following year, he dropped the wine, and introduced a syrup

Fig. 11 Advertisement of the "Vin Mariani" using popular figures of that time such as the Pope for marketing purposes

Fig. 12 Pemberton, the inventor of the refreshment beverage Coca-Cola

containing sugar and extracts of fruits from the Kola tree (which contain caffeine) and coca leaves, which was to become famous as "Coca-Cola". Mixed with water it was sold primarily as a medicine. Being billed as "The Ideal Brain Tonic" and as a remedy for headache, he in 1888 added soda water, and the drugstore soda fountain became a permanent fixture. In 1891, Asa Griggs Chandler, another pharmacist, bought all the rights to "Coca-Cola", and founded the Coca-Cola Company the following year.

The success of "Coca-Cola" spawned many competitors, among them: Wiseola, Koka-Nola, KolaAde, Celery Cola, Vani-Cola, Dr. Con's Kola. By 1903, with the many patent medicines and tonics available in the US containing extracts of coca leaves, some even having large amounts of pure cocaine, concern mounted that abuse was occurring or likely to occur.

When Atlanta introduced Prohibition in 1886, Pemberton had to replace the wine in his recipe with sugar syrup, and the wine became 'Coca-Cola: the temperance drink' (Fig. 13). Coke was soon touted as "an intellectual beverage", though not on the basis of controlled clinical trials.

Regulatories for Use of Cocaine

Due to the popularity of cocaine at its height around the turn of the century, there was bound to be some abuse. There were many state laws passed during this period to restrict the distribution of cocaine (and in a few cases opiates).

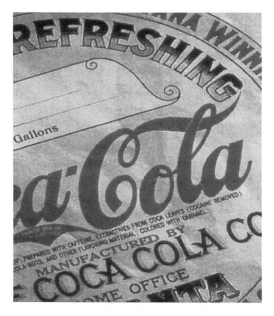

Fig. 13 Pemberton's French wine coca proved also popular with American consumers. When Atlanta introduced Prohibition in 1886, Pemberton had to replace the wine in his recipe with sugar syrup. Thus 'Pemberton's French wine cola' became 'Coca-Cola, the temperance drink'

The first restrictive law was passed in Oregon in 1887, but it wasn't until 1906 that the Federal Government attempted to regulate any drugs. In that year the Pure Food & Drug Act was passed, requiring labeling of ingredients. This effectively put many patent medicine manufacturers out of business because they didn't want the government (or in some cases the public) to know what was in their products. If, for example, they listed cocaine, they might violate certain state laws restricting or prohibiting the sale and use of cocaine. The Coca-Cola Company had wisely removed cocaine 3 years earlier to avoid such legal conflict. In 1907, the State of New York passed a very tough law against cocaine trafficking. It was launched by Al Smith, who would later become famous as a candidate for president. In 1912, 14 states required "drug education" in public schools. By 1913, 46 of the then 48 states had laws restricting cocaine, while only 29 legislated against opiates. This tends to suggest that concern over the abuse of cocaine was higher than with any other substance at that time. The Federal Government stepped in finally in 1914 with the Harrison Narcotic Act. This is the grandfather of all the subsequent drug control laws. It was originally a tax law, requiring dispensers of opiates and cocaine to register with the government, but later it was interpreted to be much broader in scope. In fact, in 1922 severe restrictions were placed on the importation of coca leaves, and this is the first time cocaine was being defined as a "narcotic", a misnomer that persists to this day.

Until 1904 the typical serving contained around 60 mg of cocaine. However, after a series of articles in East Coast newspapers calling for legal action against manufacturers of products containing cocaine, and faced with the growing opinion among the public that cocaine was dangerous, the Coca-Cola Company removed cocaine from their product. However, they continued to use "extractives from coca leaves" (i.e. E. Novogranatense var. Truxillense) but with the "cocaine removed", as shown on the label. Sold today, it still contains an extract of coca-leaves, importing 8 t a year, however with no drug in it.

Although the US Government later tried to compel the company to drop the name 'Coca-Cola'. After protracted legal argument, the name was saved; but traditionalists claimed the drink itself never quite recaptured its original glory. Although the company will neither confirm nor deny it, there is ample evidence that to this day "Coca-Cola" ("Classic") still contains de-cocanized coca leaf extracts. In fact, there is at least one reference in the literature indicating that since 1969, the Coca-Cola company hasn't even been required by the government to prove that the leaves are completely decocanized, suggesting that some actual cocaine may be allowed to slip in. The coca leaves are imported by the Stepan Chemical Company of Maywood, New Jersey, who extracts the cocaine for pharmaceutical purposes and recovers the rest of the leaf material, containing essential oils and other alkaloids, for use as a flavoring for Coca Cola.

Popularity of Cocaine as a Drug for Medical Treatment

Most medicines of the late nineteenth century with the exception of bottles with coca-extract products and/or opiates, were quite ineffective. So it's no wonder that coca extract products were considered a boon to medicine; at least they worked (Fig. 14). It also has been reported that Queen Victoria and young house-guest Winston Churchill took cocaine gum to relieve ailments whilst at Balmoral Castle. Among others, it was used as a means to stop bleeding and as a muscle relaxant. Later, in the nineteenth century it was promoted as a cure for respiratory ailments such as asthma and whooping cough. The "wonder" drug was used by many of the great minds of the nineteenth century, many of which became addicted. For instance, Oscar Wilde or John Hopkins, who discovered nerve block anesthesia, became cocaine addicts. Cocaine is also of literary interest since Sherlock Holmes supposedly took it, though it is alleged that Watson soon began to disapprove it. Unfortunately, with the cure came the problem. In 1920 cocaine was banned in the UK following reports of "crazed soldiers" in the First World War, and also due to alarming reports about the addictive potential of the drug and its ability to induce paranoia.

In the late nineteenth century more than 200 deaths were attributed to the intoxication of cocaine. In 1906, the United States population consumed as much cocaine as the US population consumed in 1976. By 1914 as crime was beginning to be attributed to cocaine usage, cocaine was started to be extracted from medicinal elixirs and Coca-Cola. As research discovered cocaine as an addictive and having the potential for abuse by users it later became illegal in the United States.

While official approval of coca-based tonics began to wane towards the end of the nineteenth century unfortunately, people who were prescribed cocaine to combat morphine dependence become addicted to both. For instance, the young Sigmund Freud (Fig. 15a), being a neuropathologist, pioneered cocaine as a treatment for postnatal depression. Though these experiments were later discontinued due to the unwanted addictive side effects, he exalted the substance, which became widely used as a panacea [2].

His "Interpretation of Dreams" and the later development of psychoanalytic method may have been aided by cocaine. Aschenbrandt (1883) and Freud (1884) were the first to describe the effect of cocaine on the central nervous system (euphoria

Fig. 14 Cocaine toothache drops in the late eighteenth century. At that time scores of patent medicines appeared, which contained extracts of coca leaves, not purified cocaine, however, some did, and lots of it

Fig. 15a Freud wrote his famous paper "On Coca" in 1884, when he was 28. In it he described the history and effects of cocaine and spoke glowingly of its therapeutic benefits

Fig. 15 (continued)

and a decrease in fatigue). In addition, cocaine was incorrectly used as a cure for morphine and alcohol abuse (Fig. 15b). Dr. Ernst von Fleischl-Marxow, who suffered from amputation (neuropathic) pain of the peripheral nerve at one hand and had become a morphine addict while attempting to ease the pain was treated by his close friend Freud with cocaine. Freud thought cocaine might cure the addiction. Soon von Fleischl-Markow, who was consuming a gram of cocaine a day, became a cocaine addict and developed a classic case of "cocaine psychosis".

Use of Cocaine as a Local Anesthetic

In 1884 Karl Koller (Fig. 16) was the first to discover cocaine to be a local anesthetic and a vasoconstrictor resulting in the first effective treatment for nasal congestion associated with seasonal allergies.

When the substance was applied to a nerve trunk, it either blocked or reversibly interrupted the passage of nervous impulses which transmit the sensation of pain to the nerve centers of the brain, and more importantly, without loss of consciousness. This peculiarity, since called the local anesthetic effect, is produced by the following mechanism (Fig. 17). The membrane of the axon (nerve trunk) is formed by a

Fig. 16 Karl Koller found that cocaine rendered the eyes insensitive to pain. A report presented on his behalf at a medical meeting in September 1884 galvanized the medical world as the only effective local anesthetic. Promoting cocaine as a local anesthetic it opened up vast new fields for surgery

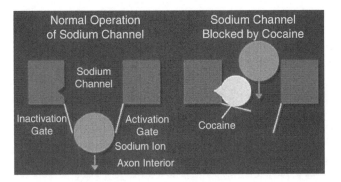

Fig. 17 Mode of action of the local anesthetic effect of cocaine

bimolecular layer of lipids, which possess hydrophilic protein layers on both sides. The local anesthetic is linked by its hydrophilic portion to the corresponding receptor of the nerve membrane and by its lipophilic portion to the other. This changes the properties of the conductor nerve membrane by blocking free flux of the sodium ions and altering the capacity of depolarization of the axon membrane, which is the mechanism through which the nerve conduction is produced. Thus, local anesthetics such as cocaine and other synthetics, inhibit the conduction of nerve impulses, which transmit the pain sensation to the brain. With greater concentrations of the drug, the sensitivity to heat and cold at first, and then the sensitivity to tact and pressure are also blocked. Very high concentrations will even impede motor impulses. The assimilation of the drug into the blood stream, however, can also produce systemic or general analgesic effects; one reason why cocaine presently is not being used as a local anesthetic any more.

Cocaine, however, boosted knowledge related to nerve function and conduction. It was Erlanger's research into nerve function with a profitable collaboration with Gasser, one of his students at the University of Wisconsin, Madison (1906–10), which culminated in profound insight into nerve function. Soon after Erlanger's appointment as professor of physiology at Washington University, St. Louis (1910–46), Gasser joined him there, and they began studying ways in which the recently developed field of electronics could be applied to physiological investigations. By 1922 they were able to amplify the electrical responses of a single nerve fibre and analyze them with a cathode-ray oscilloscope that they had developed using different agents for blockade. The characteristic wave pattern of an impulse generated in a stimulated nerve fibre, once amplified, could then be seen on the screen and the components of the nerve's response studied. In 1932 Erlanger and Gasser found that the fibres of a nerve conduct impulses at different rates, depending on the thickness of the fibre, and that each fibre has a different threshold of excitability – i.e., each requires a stimulus of different intensity to create an impulse. They also found that different fibres transmit different kinds of impulses, represented by different types of waves. As early as 1929 the two physiologists Erlanger and Gasser experimented with pressure to evaluate the capacity of cocaine to block nerve conduction in frogs.

Thanks to its local anesthetic property, it is possible to apply a 2% cocaine solution on the nerve of a diseased molar and remove it without having the patient suffer and be tortured by pain. The doctor is able to find a calm and passive subject on whom he could work at ease and carefully. One can only imagine that the benefits were momentous in the history of buccal surgery. It marked the passing of the traumatic, painful, dangerous and primitive surgical methods to the painless surgery of the twentieth century, which permitted great advances in the medical sciences. The coca leaves and the miraculous substance against pain, cocaine, soon rose to the pinnacle of pharmacology and medicine, which most of all showed no development of tolerance [3]. For instance, a 2% cocaine solution for odontological and ophthalmological work, as well as ocular surgery became a reality (Fig. 18). The anesthetic properties of cocaine were used to an advantage for making medications against birth pains, ointments for hemorrhoids, solutions to relieve dentition pains in infants, drops for earaches, in addition to the myriad of applications in all surgical cases of the various medical specialties: traumatology, abdominal surgery, gynecology, etc. (Fig. 19).

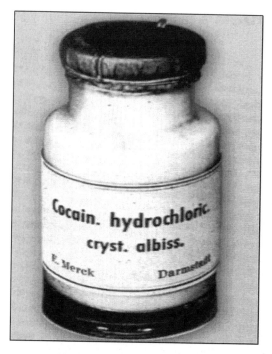

Fig. 18 Cocaine hydrochloride the basic salt as it was used in pharmacy as a base for various preparations

---

**NEW MAGISTRAL FORMULARY
OF CLINICAL AND PHARMACOLOGIC
THERAPEUTICS
by
DOCTOR ODILLON MARTIN**

(Former Chief Laboratorist of the School of Medicine of Paris)

EYE DROPS
    Cocaine Chloride.. 10 cg
    Distilled Water .... 10 g

INTRA RACHIDIAL INJECTIONS
    Cocaine Chloride.. 10 cg
    Boiled Distilled Water .... 20 cc

Sterilize: the solution to be applied must always be recently prepared, inject into the preferred site (See dispensing methods) 2 or 3 cc of the cocaine solution. Preparatory surgical anesthetic.

OINTMENTS
    Alcohol coca extract... 1 g
    Benzoate pork fat or starch glycero .... 15 g
    To relieve continuous painful locations, fissures, sores, etc.

SUPPOSITORY
    Alcohol coca extract... 1 g
    Cocoa Butter .... 4 g

To relive the pain of anal fissures and hemorrhoids

Fig. 19 Few examples of the vast assortment of cocaine preparations being used in medicine, ca 1910

The Different Forms of Cocaine

The drug cocaine comes in five different forms

1. **Coca Leaves**: The cocaine content in the coca leaf is only 0.1–0.8%. The plants in the higher altitudes tend to have a higher cocaine content then those at lower altitudes. There are different styles in using the leaves to extract the stimulant. The leaves can be rolled into cigarettes or cigars and smoked, the leaves can be infused with a liquid to form a tea or they are chewed.
2. **Coca Paste**: The paste is the middle step between the leaf and the powder cocaine. It cannot be injected or snorted, therefore the only use of the coca paste is to burn the substance and inhale. Coca paste is much more popular in South American countries then in the United States.
3. **Powder Cocaine**: Powder cocaine in general, is derived from dissolving coca paste with hydrochloric acid. Powder cocaine is the most widely used form of cocaine, and it is cocaine at its purest form. Usually street cocaine is not necessarily pure as it is mixed with different additives such as sugars, local anesthetics, or other drugs. Powder cocaine can be snorted, injected, or ingested. Because it decomposes easily at temperatures above 198°C or 388°F, unlike crack cocaine the powder cocaine cannot be smoked. Powder cocaine loses its potency when the drug is heated. Because the cocaine alkaloid decomposes at high temperatures it does not produce the physiological or psychotropic effects.
4. **Cocaine Base**: Cocaine base is obtained from powder cocaine, which chemically is cocaine hydrochloride and is formed by reacting with a base at controlled conditions. It resembles much the crude cocaine paste, however, having a higher purity. At the melting point of 98°C the cocaine base loses its mind altering effects. The cocaine base is poorly absorbed throughout the body and thus does not carry a similar level of potency of powder cocaine.
5. **Crack Cocaine**: Crack results from heating powder cocaine with, for instance baking soda. By this reaction the hydrochloride acid escapes from the salt while carbon dioxide is formed from baking soda. Since this reaction results in crack-like noises, the final constituent is called "crack", which is more volatile at low temperatures. Thus crack cocaine allows the user to smoke the cocaine instead of injection or insufflation. A crack cocaine rock can be anywhere between 75–90% pure.

The Different Types of Alkaloids in Coca

Coca contains at least 14 separate alkaloids. An alkaloid is a naturally occurring nitrogen-containing compound, which shows a basic, i.e. an alkaline reaction, resulting in water-soluble salts when added to an acid. It is assumed that alkaloids are produced by the plant as a type of defense against insects and herbivores. They are usually bitter tasting, and often have psychoactive properties. The most important alkaloids of the coca plant are as follows:

1. **Cocaine**: (benzoylmethylecgonine) The principal alkaloid, which can account for between 20% and 90% of the total alkaloidal content of a leaf. The E. Coca or Huanaco (Bolivian) leaf usually contains the highest amount, and is the most prized.
2. **Cocamine**: Found in high concentrations (up to 80%) in E. Novogranatense, especially the var. Truxillense (Trujillo leaf).
3. **Cinnamylcocaine**: Found in E. Novogranatense, especially in the Java variety.
4. **Hygrines**: These are agents with properties similar of an oil and therefore are useful as flavoring agents. Found mostly in E. Novogranatense, they probably account for the value of this leaf as a beverage flavoring.
5. **Benzoylecgonine**: Most likely a decomposition product formed during the breakdown of cocaine.
6. **Tropacocaine**: Found only in significant amounts in E. Novogranatense, a Java variety.
7. **Ecgonine**: Again, probably a decomposition product.

Up to this point, only coca has been discussed. However, it must be considered separately from cocaine due to its being used in its entirety (as the whole leaf) by the indigenous population of the Andes mountains. An attempt to extrapolate data from those who use the whole plant to those who use only the purified principal alkaloid (cocaine) is meaningless. Thus there is a paraphrase which seems perfectly correct: " *When one removes the substance from its protective green envelope, the problems begin*".

The Making of Cocaine in the Jungle

All of the illicit cocaine entering the US and Europe comes from South America where it is processed in makeshift jungle laboratories. These are located close to growing areas because the leaves are hard to transport far, owing to their weight (it takes 200–300 lb of leaves to make a kilo of cocaine) and the inability to conceal them easily. In contrast to the common laboratories with clean buildings, a lot of expensive glassware and doctors with white lab coats on, most cocaine labs look a lot different. They are equipped with a few plastic buckets, some acid, some alkali, and one or two cooks ("cocineros"), the underground chemists "all they need to make cocaine is a couple buckets and a sheet". These labs are only in operation for a short period of time (2 weeks or less) two or three times a year during the harvest.

These labs are meant to extract the alkaloids from the coca leaf, which yields a mixture of several alkaloids and oils from the leaf. This process is usually carried out at or near the growing areas. With a yield of about 1% cocaine per leaf, it takes 200–300 lb of leaves to make a kilo of crude cocaine (Table 1). The leaves are put in a pit or drum and mixed with calcium carbonate, then left to stand in the sun for a few hours so that the calcium carbonate will "soften" the leaves and cause them to "sweat" their alkaloids out. This process is known as "La Salada", the "salting". The leaves are then soaked ("La Mojadura", the "soaking") in kerosene (or gasoline if that's what's available), which will dissolve the alkaloids. The leaves are then pressed to recover the kerosene, known as "La Prensa" the "pressing", and thereafter are discarded (Fig. 20). Water, with diluted sulphuric acid is added to the kerosene, and since the mixture is now acidic, the alkaloids precipitate from the kerosene, and they will dissolve in water, a process called "La Guaraperia", or the "separation". The kerosene is now poured off, and the water (now containing the dissolved alkaloids) is made basic by adding an alkali (ammonia). This will cause the alkaloids to precipitate, after which they are filtered usually with a sheet stretched between four sticks, wrung dry and left under a lightbulb or the sun to dry thoroughly. This process is called "La Secaderia", or the "drying". The resultant product is known as "Coca Paste" or "Pasta", the alkaloidal cocaine. However, this alkaloidal cocaine should not be confused with cocaine base, because being contaminated with other by-products it is not as pure as the high quality cocaine base, which is yielded by the next step in the refining process.

Table 1 The table outlines the first of three steps usually taken in the refining of illicit cocaine. This step is directed at obtaining "Coca Paste" also known as "Pasta"

- *Coca leaves + lime (carbonate) are soaked in kerosene (24–36 h)*
- *Leaves are separated (pressed) from kerosene & discarded*
- H_2O + *diluted sulphuric acid is added to kerosene which will results in the precipitation of Coca alkaloids from kerosene and dissolving in the water*
- *The water (containing the alkaloids) is separated from kerosene*
- *Alkali (ammonium hydroxide) is added to the water in order to precipitate alkaloids*
- *The precipitate is filtered and dried in the open, which eventually will lead to the valuable Coca Paste – "Pasta"*

Fig. 20 A "cocinero" pressing the soaked leaves to separate out the kerosene containing the alkaloids ("La Prensa") in a typical cocaine jungle laboratory

Steps Refining Coca Paste to Cocaine Base

Following harvest of the leaves and the steps to get crude cocaine paste, is the refining process to yield cocaine base. The refining process of illicit cocaine is quite different from the process used to make pharmaceutical-quality drug. There, in the ecgonine conversion process for making medicinal cocaine, the total alkaloid content is extracted from coca leaves, usually E. javanese. This extract is saponified, yielding ecgonine as its main product, which then is refined by recristallization. The resulting high-grade ecgonin then is reacted with benzoyl hydrochloride and with methanol and thus is converted to cocaine.

The illicit process, however, tries to preserve the naturally occurring cocaine and avoid its degradation into ecgonine. The purification is performed through the steps extraction, precipitation, solution and again precipitation, as outlined above. And although the illicit chemist ("cocinero") attempts to exclude all other alkaloids from the final product, this never happens completely. With at least 14 alkaloids found in coca leaves, there will be some (at least trace amounts) present in the resulting cocaine. Hence it is rare to find illicit cocaine that is more than 95% pure. It must also be remembered that the residues of chemicals such as benzene, acetone, sulphuric and other acids and even kerosene and gasoline and their additives will often be identified as impurities.

Coca paste or "Pasta" is the result of the first step in the refining process, As mentioned before, this is a mixture of alkaloids including cocaine with a beige color. If the leaves used to make the "Pasta" were of exceptionally high quality (Huanaco or Bolivian leaves) with a 90% cocaine to other alkaloids content, this "Pasta" might not have to be further refined into cocaine base, as it would already be 90% cocaine anyway. In that case, the chemists would simply add hydrochloric acid to the "Pasta", which would result in the finished product – cocaine hydrochloride or street coke. However, this is rare, as most illicit labs don't get leaves that are of high quality, so they must perform another step in the refining process to obtain cocaine base. It should be noted that while much of the "Pasta" produced is eventually converted into cocaine base and then to cocaine hydrochloride, a fair amount of this Coca Paste is diverted for local consumption. As has been stated already, the melting point of the cocaine base is 98°C, while cocaine hydrochloride melts at 190°C. It is because of this difference in the melting point that cocaine base is suitable for smoking (because it sublimes easily) a process known as "Basuko". This has become a tremendous public health problem in the producing countries themselves, accounting for severe dependence and side effects – due not only to the cocaine, but to the contaminants of the manufacturing process, especially lead from kerosene and gasoline additives. A second step is usually necessary, because the coca leaves used to make the Coca Paste, or "Pasta" contain varying amounts of cocaine and other alkaloids. This process is designed to separate the cocaine from the other alkaloids in the mixture. Coca Paste is added to diluted sulphuric acid and

Table 2 The different steps in refining coca to cocaine base

- *Coca paste ("pasta") added to water + diluted sulphuric acid*
- *Potassium permanganate is added to remove inessential alkaloids through oxidation*
- *Alkali (ammonium hydroxide) is added to stop the action of permanganate and precipitate the cocaine base*
- *Filter dry, which eventually results in the Cocaine Base*

Table 3 The final steps in refining cocaine to street cocaine

- *Cocaine base dissolved in ether + acetone*
- *Hydrochloric acid (HCl) is added to precipitate the crystals, filtered and dried, and the result is*
- *Cocaine Hydrochloride*

water upon which the alkaloidal "Pasta" will dissolve. Potassium permanganate is added to decompose out inessential alkaloids and other by-products through oxidation (Table 2). This step is critical, and if not done carefully, can result in the destruction of the cocaine as well as the unwanted alkaloids. If this step is omitted, the resultant base is rarely more than 65% pure cocaine. Once oxidation is complete (signalled by color changes), and alkali (ammonia) is added to stop the action of the permanganate after which cocaine base precipitates, then being filtered and dried. This process can be carried out almost anywhere, since there aren't any leaves to carry around. Often, such "base" labs are found in the jungles of Colombia, and Coca Paste is brought to them from several "Pasta" labs located near growing areas throughout the Andes region.

The third step in refining is to convert the Cocaine base into its salt (Table 3). This final step in manufacturing illicit cocaine is the making of "Crystal", which is the South American name for cocaine hydrochloride – not to be confused with methamphetamine which is often called "crystal" in the US. The hydrochloride (salt) form of cocaine is the most common form sold in the street. The primary reasons for its use as a salt is the fact that the hydrochloride is soluble in water, so it can be applied to mucous membranes (snorted), or put into solution and injected. In addition, it is also more stable for storage purposes than the base form. The following process can be carried out almost anywhere, often an apartment in Bogota, Colombia, is all that's required: Cocaine base is dissolved in ether and acetone. Hydrochloric acid (HCl) is added, and this will precipitate crystals of cocaine hydrochloride, which is not soluble in organic solvents like ether. The crystals are filtered off and dried. This product is cocaine hydrochloride or street coke (Fig. 21). As will be seen later, some users are converting their cocaine hydrochloride back into cocaine base, by chemically "freeing" the base from its hydrochloric salt in order to get a product that can be smoked.

Fig. 21 Appearance of "Crystal", or street coke; this is what the final product is supposed to look like

Trafficking of Cocaine: The Legal Approach

Once the cocaine has been legally produced from the coca leaf, it is exported to various countries for medicinal use, basically as a topical local anesthetic (applied to the surface, not injected, only treating a particular area). In the United States the crystalline powder is imported to pharmaceutical companies who process and package the cocaine for medical use. Merck Pharmaceutical Company and Mallinckrodt Chemical Works distribute cocaine in crystalline form (hydrochloride salt) in dark colored glass bottles to pharmacies and hospitals throughout the United States. Cocaine, in the alkaloid form (base drug containing no additives such as hydrochloride in the crystalline form) is rarely used for medicinal purposes. Cocaine hydrochloride crystals or flakes come in 1/8, 1/4 and 1 oz bottles from the manufacturer and has a wholesale price of approximately $20-$25/oz (100% pure).

Cocaine is still a drug of choice among many physicians as a topical local anesthetic because the drug has vasoconstrictive qualities as it stops the flow of blood oozing. And although synthetic local anesthetics such as novacaine and xylocaine (lidocaine) have been discovered and are used extensively as local anesthetics, they do not have the same vasocontrictive effects as cocaine.

Profit Making and Trafficking of Cocaine: The Illegal Approach

The hierarchy of cocaine traffic is as follows:

1. Cultivation, harvesting and selling of dried coca leaves to clandestine labs at a price from $1 to $3/kg.

2. Extraction of cocaine from the dried coca leaves and selling pure cocaine to the wholesaler at a price of $3,000/kg realizing a profit of approximately $2,500/kg.
3. Directly selling or smuggling the kilo of pure cocaine to the distributor for a price of $18,000/kg, realizing a profit of $15,000/kg.
4. *Cutting* the cocaine and selling it at pound quantities to dealers for $10,000/lb (50% pure cocaine) realizing a profit of $22,000 from the original kilo he purchased.
5. *Cutting* the cocaine and selling it at ounce quantities (30–40% pure cocaine) to the pusher for $800-$1,200/oz, realizing a profit of $10,000 from the pound the pusher originally purchased.
6. Possible *cutting* the cocaine and selling it by the gram (spoon) quantities to users for $50-$75/g, realizing a profit of $1,000 from the original ounce the pusher has purchased.

The original 100 kg of dried coca leaf that it takes to produce 1 kg of pure cocaine costs approximately $200. The kilo of pure cocaine will eventually be worth over $200,000 when sold to users in 25% pure gram quantities.

The illicitly manufactured cocaine from the various clandestine cocaine labs in South America, is smuggled to various countries including the United States for black market trafficking and use. Those involved in the smuggling of cocaine vary from a one-man operation to organized syndicates. The smuggling methods are unlimited and vary with one's imagination. Often times, cocaine is first smuggled into Mexico rather than directly from South America to the United States. Cocaine, as is heroin, is usually packaged in hermetically sealed plastic bags or rubber condoms for smuggling purposes. Once the cocaine enters the US, it is then distributed through various sub-dealers down to the users. Illicit cocaine, basically, comes in three forms:

1. The hard tiny rock form, which is readily available, especially to the large wholesaler or dealer.
2. The flake form, which is generally fairly pure cocaine, which has been broken down into tiny flakes and considered a delicacy among users of cocaine.
3. The powdered form, which is usually rock or flaked cocaine diluted with other substances such as lactose or procaine.

In the illicit traffic of cocaine, as in many other drugs, there is a definite channel, which the drug goes through from the harvester to the user. Initially, there is the farmer who cultivates, dries and ships the coca leaf to the illicit lab. The clandestine labs then chemically extract 90–100% pure cocaine from the leaf. From the lab, the cocaine is usually sold to smugglers or wholesalers at a price of $200 an ounce or $3,000 a kilo. The wholesaler smuggles the cocaine into the United States and sells it to a major cocaine distributor for a certain agreed upon price, which varies and ranges from $18,000 to $22,000 a kilo. The distributor will then take the large quantity of cocaine and sell lesser amounts to a number of dealers. He may sell the cocaine in its pure form or dilute it and sell more for a lower price. Most of the traffickers keep in mind that cocaine loses its strength readily and sometimes the cutting or diluting agent will have a tendency after a period of time to begin destroying the cocaine content. When the dealers are in possession of their 0.5 lb or pound of cocaine, they will most often dilute it with a cutting agent and sell it in ounce quantities to even smaller dealers. The cocaine street pusher will in turn, sell it to the user in gram quantities.

Smuggling of Cocaine

Transportation of refined cocaine to the end-user, apart from private planes and small fisher boats, also commercial airlines are used extensively for smuggling. For example Avianca, the national airline of Colombia was often accused of carrying cocaine. Once, a US carrier (Eastern Airlines) jetliner was confiscated by the US Customs as a warning to stop letting their airplanes be used to ferry cocaine.

Everything has been tried to smuggle cocaine into the Western countries. For instance, envelopes that had been soaked in cocaine in Colombia, were mailed to the US where the cocaine would be recovered. And often so-called "body packers" do the smuggling (Fig. 22). The "body packers" are individuals who swallow cocaine carefully wrapped up in condoms or double latex balloons in order to get through customs. Arrived at their destination, they then pass the packages, taking careful note of their number. This is not always uneventful, and there have been several instances of sudden deaths in the customs inspection area.

To get some idea as to the size and number of "packages" that can be carried the following figure (Fig. 22) shows an X-ray of the abdomen of a subject taken recovering over 90 (!) bags.

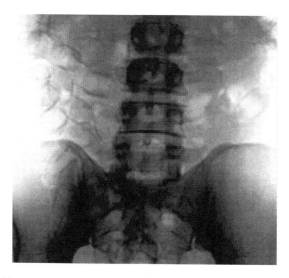

Fig. 22 Bags with cocaine salt before being swallowed and an X-ray of a body packer smuggling cocaine

Diluting or "Cutting" of Cocaine

Rarely does anyone, except high echelon dealers, come in contact with pure cocaine. All along the distribution chain the drug is cut or "stepped on". Adulterants are added to increase weight, and then other, mainly less costly, psychoactive substances are added to make up for lost potency. In the case of cocaine, a typical type of cut is often found: an "anesthetic" cut. This is usually procaine or lidocaine or just about any other synthetic local anesthetic. It really doesn't matter because the reason is to provide a "freeze" for the user when he either tests the product by dabbing some on the tongue, or snorts it. The resultant "freeze" is usually enough to satisfy the buyer that the coke is of *good quality*. The major problem with cuts is the added toxicity they confer on the primary drug. The greatest toxicity comes from "active" cuts such as phencyclidine (PCP), amphetamines or, what is frequently seen, phenylpropanolamine. Commonly found in cold medications and over-the-drug counter (OTC) diet capsules, phenylpropanolamine can augment the cardiovascular effects of cocaine, and its effect lasts longer. Another "active" cut is yohimbine, a central alpha-2 blocker with effects opposite to clonidine, which is an alpha-2 agonist. The former drug causes increases in blood pressure and heart rate as well as increased motor activity. It is abused as a general stimulant, and has been illicitly marketed as an aphrodisiac. The diluting agents for the cutting of cocaine are very similar to those used for heroin. One of the basic differences between "stepping on" (diluting) cocaine as compared to heroin, is that cocaine is usually only diluted down from 20% to 40%. The process for cutting cocaine varies from individual to individual. Often times the large dealer use a more elaborate process, but the basic operation is the same throughout cocaine traffic. The cutting or diluting agent used for cocaine again varies with the individual and the substance that is readily available to that individual. Some of the common cutting agents for cocaine are see in Table 4.

Table 4 Summary of cocaine adulterants, so called "cuts", when it comes to sell cocaine on the streets

Inert drugs	Pschoactive drugs	Local anesthetics
Mannitol	Amphetamine	Lidocaine
Lactulose	Methamphetamine	Procaine
Sucrose	Phenylpropanolamine	Benzocaine
Maltose	Phencyclidine	Tetracaine
Inositol	Quinine	Bupivacaine
Flour	Yohimbine	
Corn starch	Caffeine	Meaverine
Talcum powder	Guarana	Mepivacaine
Ascorbic acid	Strychnine	Articaine
Citric acid		Prilocaine
Plaster	Colchicine	Ropivacaine
Sodium bicarbonate	Acetaminophene	
Sodium borate (Borax)	Ketamine	
Sodium chloride		

Procaine and other local anesthetics present synthetic preparations in powder form, while *Mannite* or *Mannitol* is a sugar substance used as a laxative and produced in Italy. *Menita* is a milk sugar from Mexico and South America, and *Lactose* or *Dextrose* present white powdered milk sugar used as a baby food supplement, which can be purchased readily in any drug store.

Another more active cut is powdered *methamphetamine* also known as "speed", while *quinine* is used to treat leg cramps and malaria. Many times powder *vitamins* purchased in health food stores and just about any soluble powder that is not disruptive to the body is used. Some typical examples are *baking soda, powdered sugar, powdered milk, starch*, or the laxative *Epsom salts*, a chemical compound containing magnesium sulfate, often encountered as the heptahydrate. The dealer will either be told the percentage of cocaine by a trusted "connection" or he will be able to approximate the percentage by various means.

Ascertaining the Purity of Cocaine and the Cutting Agent

There are several ways how to ascertain the quality of the merchandise

1. **Quantitative chemical analysis**, which is an elaborate process requiring a qualified chemist and special laboratory equipment.
2. **Cocaine drug testing kits** either manufactured for law enforcement purposes or produced by the underground. These testing kits are simply presumptive color tests. The basic color test used for cocaine is *cobalt thiocyanate*. The cocaine or any of the other substances from the "caine" family will form a brilliant blue flaky precipitate with cobalt thiocyanate. This is an indication that the product is cocaine, procaine, tetracaine, etc. In order to determine whether there is actually any cocaine and not all procaine, *stannous chloride* is added to the precipitate causing all of the caines except cocaine to dissolve. If the dealer suspects that cocaine has been cut with a local anesthetic, he can then make a partial determination as to how much of the procaine or another "caine" is contained in the total powder by a more selective test.
3. **The chlorox test**. Chlorox is a brand bleach consisting of a 5.25% solution of sodium hypochlorite and is typically used as a household bleach. It is alleged that the dealer can take suspected cocaine and drop it in a vial of chlorox. While the cocaine will dissolve completely, procaine (or any other local anesthetic) will turn a reddish orange color with trailing to the bottom of the vial as residue. Although the test is very unspecific, it still is used by many dealers.
4. **The water test**. It is also alleged by the street dealers, that a determination can be made as to how much cut is in the cocaine by placing the powdered substance in a glass of water. Although not very meaningful, allegedly cocaine will dissolve almost immediately leaving the remaining cut which normally will dissolve slower and not as clear.

5. **The burning test**. The powdered cocaine is placed on aluminum foil and held over a low flame or match. The cocaine will burn clear. A sugar cut will darken and burn a dark brown or black therefore the larger the cut, the darker the burn. Crystallized speed or methamphetamine will pop when burned. Salts do not burn and remain as residue (cuts such as procaine or quinine also burn fairly pure although it is alleged that procaine can be detected by a bubbling of the substance before it burns clear).
6. **The methanol test**. Most common cuts do not dissolve in pure methanol although cocaine does. Unfortunately for the dealer, procaine and methamphetamine also dissolve in pure methanol. It is imperative that pure methanol be used since any water in the alcohol will tend to dissolve other cuts such as sugar and salt. Methanol can be obtained in most paint supply stores. The dealer will take two equal amounts of the cocaine substance and place the equal amounts in two teaspoons next to one another. Then a quarter of a teaspoon of pure methanol is added to one of the spoons. The mixture is then stirred and any powder that remains is compared to the original unaltered amount in the second teaspoon to determine the percentage of the cut. If, for example, 20% of the original amount did not dissolve, the substance tested would be no more than 80% pure.
7. **The sodium carbonate test**. If the suspected cut is procaine, the cocaine substance can be added to a sodium carbonate solution. This would dissolve all the cocaine leaving just the procaine.
8. **The use test**. Some dealers will test the percentage of cocaine by snorting (Fig. 23). This is probably the best and most common street test in determining the purity of the cocaine. The tester should standardize the amount snorted so that he will have the ability to distinguish. The tester will look for the swiftness of the high and the "freeze" or numbness the substance causes. If the nasal passages

Fig. 23 Snorting of cocaine, a common practice for testing of cocaine quality but also a method of abuse

burn and the eyes tear, there is a good possibility the cocaine has been cut with speed. Sugar and salt cuts quite often cause a post nasal drip. Excessive sweating and hyperactivity could mean either a speed or quinine cut was used. Excessive diarrhea would denote a laxative type cut such as epsom salts, or menita. Speed tends to cause irregular bowel movement. A greater degree of numbness indicates the presence of procaine or other local anesthetics.

9. **The taste test**. Cocaine has a bitter taste and the added cut will tend to alter that taste. A milk sugar cut will sweeten the cocaine although dextrose has a tendency to sweeten the substance more than lactose. Procaine will be bitter to the taste but will tend to numb the gums and tongue quicker and longer than cocaine. Sodium chloride has an after taste and Epsom salts are bitter in taste and sandy in texture.

10. **The observation test**. Pure cocaine crystals have a shiny almost transparent appearance and even when crushed, will retain the crystalline sparkle. The crystalline sparkle of cocaine will be dulled by most cuts. Dextrose has less dulling effect than lactose although a speed cut usually dulls the crystals less than most other cuts. Although salts have a crystalline structure, they tend to be duller than the cocaine crystals. An alleged indication of the purity of the cocaine is also the tiny rock-like material contained within a probe. The tiny rocks are most likely pure cocaine as they come from the manufacturer. The rock or hard substance can be ascertained by feeling the powdered substance.

Once the dealer has assured the purity of the cocaine and/or the cut or adulterating agent, he is then ready to begin the process of "stepping on" the cocaine. Most dealers will dilute a small portion of the cocaine and then re-test it. Many dealers claim that they usually only cut the amount of cocaine that will immediately be sold because the cuts have a tendency to destroy the stability of cocaine. It is therefore advantageous for the dealer to keep the cocaine sealed in a cool place such as the refrigerator and in an amber or dark-colored jar to retain the strength of the drug as long as possible. Dealers claim that with time, moisture, warmth, air and sunlight decrease the potency of the cocaine.

The process for *"stepping on coke"* again varies with individuals, but the two basic formulas are similar to those of heroin and are as follows:

- 1 oz of lactose added to 1 oz of 100% cocaine = 2 oz of 50% cocaine
- 2 oz of lactose added to 2 oz of 50% cocaine = 4 oz of 25% cocaine
- 2 oz of lactose added to 1 oz of 100% cocaine = 3 oz of 33.3% cocaine
- 3 oz of lactose added to 1 oz of 100% cocaine = 4 oz of 25% cocaine
- 4 oz of lactose added to 1 oz of 100% cocaine = 5 oz 20% cocaine

The dealer will measure off the desired amount of cocaine, for instance five level teaspoons, and place it in a pile on a flat nonporous surface such as a record album, mirror or glass plate. He will then measure off the desired amount of lactose and place it in a separate pile on the same surface. Then, using a playing card, razor blade, knife or any sharp edged instrument, the dealer chops the cocaine to take out all lumps so the cocaine is a fairly fine powder. In order to separate any foreign material, the mixture is then sifted through a sifter or nylon stocking so that a fine fluffed powder is obtained. Once through the sifter, the cocaine usually has a little more volume since it has been fluffed. The cocaine is then sifted into a pile and the same process is repeated with the diluting agent. Thereafter, the dealer will mix the pile of cocaine into the pile of diluting agent. Once this has been accomplished, he sifts the diluted cocaine through a sifter trying to get the mixture as equal as possible. At times the dealer will resift the diluted cocaine to assure an equally distributed mixture, which then is ready to be placed into packages for sale. Once the dealer has diluted the cocaine, he will measure off the desired amount to be packaged by weighing it on scales.

Economics of Cocaine

This is a rather simplified attempt to trace a 100 kg (220 lb) shipment of cocaine from the growers to the users (Fig. 24). At this point it might be useful to say a few words about the organization of the business. The major traffickers are thought to

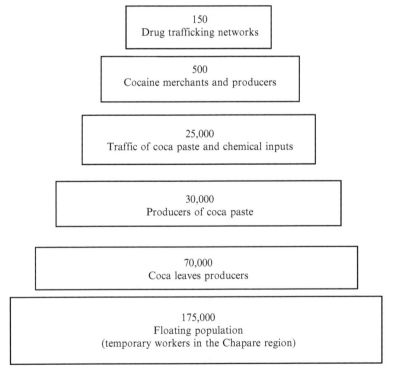

Fig. 24 Representative example of the economics of cocaine in Bolivia of a population employed in coca and cocaine production

Economics of Cocaine

be a collection of 12 South American families, who control the cocaine trade at every level, from grower to consumer (vertical integration). The chart shows the number of people involved at various levels. Note that the one point, where everything comes down to a single person, is at the level of the smuggler ("mule"), and it's no wonder that interdiction efforts are concentrated at this point.

The estimate of monetary value is based on 1 g of street coke, purity around 30%, and costing the user US$100. Naturally, prices fluctuate, especially at the present writing, when there is a glut of cocaine on the market and prices are falling, but there's still a big "sucker" trade out there, and not all users are getting high quality cocaine at US$65/g. The current estimate of the size of the illicit market is US$80 billion/year for all drugs.

Freebase Cocaine: High Bioavailability with Increase in Potency

During the late 1960s, American cocaine smugglers travelling in South America noticed some of the local populace of Peru and Colombia smoking a substance identified to them as cocaine. When word of this reached the US, several users experimented with smoking cocaine hydrochloride sprinkled on cigarettes, and quickly discovered that street coke literally goes up in flames. What was wrong? The smugglers had disregarded an important point: the Peruvians were smoking Coca Paste ("Pasta"), which contains alkaloidal cocaine (cocaine base). While cocaine base melts already at about 98°C, cocaine hydrochloride (street cocaine) melts at 195°C with decomposition. The task then was to find a way to perform a little "reverse chemistry", that is to turn cocaine hydrochloride back into cocaine base. The process to accomplish this was already in use by some skilled users who wanted to "purify" their cocaine by turning it into the base in order to remove diluents and cuts. Then they promptly turned that base back into cocaine hydrochloride. It was when someone put two-and-two together that cocaine "freebase" was born.

Freebasing in the first place is the creation of smokeable cocaine, and the only difference between the base and its salt is that the hydrochloric acid is removed in the base form. To put it in other words, the cocaine base is "freed" from the hydrochloric acid, hence the term "free base". Its really just alkaloidal cocaine, or cocaine base, the step in the illicit refining process just before the production of "crystal". The base form is soluble in ether but not in water. The salt form is soluble in water but not in ether. The base form is smokable but not snortable or injectable. The salt form is snortable or injectable but not smokable. The difference between the base and the salt is also seen in the onset and the intensity of effects. Thus, taking cocaine orally produces effects within 15 min, and intranasally ("snorting") within 3–5 min. An intravenous administration, however, will kick-in within 30–45 s, while smoking (cocaine base) takes only 7–10 s (Fig. 25).

There are some major hazards associated with "freebasing" that go beyond the mere prospect of blowing oneself up. Note the rapidity with which an effect is achieved, and the short duration of that effect (Fig. 26). It is these two things that make "basing" the most dangerous and addictive form of cocaine. Therein lies the reason for "freebasing", the quest for the ultimate "high". The method of producing "freebase" (such as the ether method, see next page) is useful only if one desires to

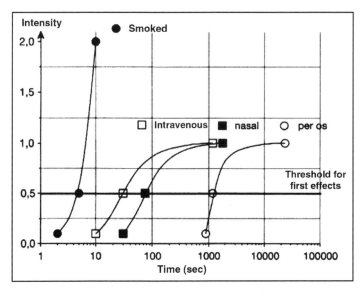

Fig. 25 Differences in intensity, onset and duration of action of cocaine hydrochloride as it is used in snorting versus the cocaine base as it is used in smoking

remove some of the inert non-baseable cuts such as mannitol or lactose, but an unnecessary, expensive and dangerous procedure. Quickly, dealers came up with a way to provide "pre-based" cocaine to their hungry customers in the form of "Rock cocaine" or what is known in the East as "Crack". There are several procedures that will accomplish this.

Freebasing of Cocaine: The Ether Wash Method

There are several ways that "freebase" is produced. The ether wash method is the original procedure. It's really fairly simple acid/base chemistry, but and most dangerous to perform (Table 5). First cocaine hydrochloride is dissolved in water, which results in a weak acidic solution. An alkali such as ammonia or bicarbonate is added upon which a precipitate is formed being insoluble in the watery solution. When ether is added to the water it will first float on top. When shaken the precipitate now dissolves completely in the ether phase. The two layers, ether and water, are now separated in a separate funnel, or if only small amounts are needed, the upper (ether) layer can be removed (usually with an eye dropper) and transferred to a petri dish where it will rapidly evaporate leaving behind the base crystals (Fig. 27).

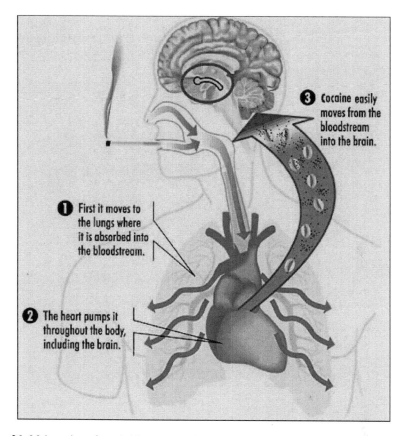

Fig. 26 Main action of smokable agents such as nicotine and crack cocaine. Due to the high lipophilicity there is a fast diffusion into the blood stream, followed by a rapid onset of action and an increased intensity of effects

Table 5 Cocaine freebasing; the "Ether Wash Method"

- *Cocaine hydrochloride + water*
- *Alkali is added (ammonium hydroxide) and the base will precipitate out*
- *Ether is added, and the base dissolves in the ether phase*
- *Ether is separated by evaporation, the residue is*
- *Cocaine "Free" Base*

As seen in the following Fig. 28, a special base pipe is lit where the cocaine is put into the receptacle at the top and heat applied, either from a cotton ball dipped in rum, or a hand-held torch as shown. If the ether extraction method of producing base has been used, this co-existence of flame and flammable solvent can be quite illuminating. Working with ether is like playing with liquid dynamite. Therefore, in order to avoid an explosion hazard, it is essential to keep away any fire in the room or the surroundings where ether is being used. The base, when properly dried (freed from ether), is not combustible.

Fig. 27 The difference in appearance of cocaine hydrochloride and its base as crystals (*right*). This is the finished product after all the ether had evaporated

Fig. 28 A freebaser with his paraphernalia lighting up a glass pipe in which there are several stainless steel mesh screens to trap the melting cocaine base as it drops into the pipe after heat is applied. The base will vaporize in time, and this is inhaled for a rapid effect, and an equally rapid "crash"

Because of the flammability of ether, there is a big problem with this method, and a few myths. Users think that they're basing out some of the impurities in the cocaine, not so. Inert cuts such as mannitol, lactose etc. are indeed removed, but most of the active and anesthetic cuts are baseable. In short, they base right out with the coke. Thus, smoking these active cuts along with cocaine might increase the danger of an already toxic agent considerably.

Freebasing of Cocaine: The Baking Soda Method

With the obvious problems of using ether to produce free-base cocaine, its popularity declined precipitously after repetitive near-fatal burning in fires after freebasing binge. Street chemists went back to their drawing boards trying to find another solution for freeing the cocaine molecule from the hydrochloride tag (Fig. 29). It all came down to one feature that was the real reason for freebasing in the first place, i.e. the creation of smokable cocaine. Now the most popular method involves the use of a common household alkali such as sodium bicarbonate or baking soda. Another is ammonium hydroxide or ammonia. This is mixed with the cocaine hydrochloride (street coke). When being heated sodium chloride is formed and bicarbonate dissolves into CO_2 and H_2O. Freebase cocaine not only contains all the original cuts found in the street coke, but now it also is adulterated with baking soda. It is a marketer's dream, to take an adulterated product, add more fillers and sell the resultant substance in small pieces for a profit that quadruple the cost of the original product. At the same time, the customers come back more often because the product's effect wears off much sooner than if it were the original substance street coke. Why not sell cocaine hydrochloride for $100/g, when that same gram turned into "Crack" will sell for $300-$400 broken up into little pieces?

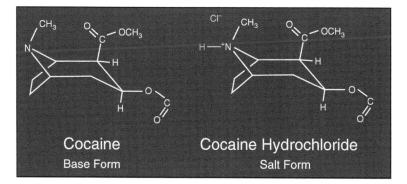

Fig. 29 The difference in the basic molecule structure of cocaine base and cocaine hydrochloride, which results in different physicochemical and addictive properties. For better viewing the aromatic ring benzoic acid is not shown

Pharmacology of Cocaine

Chemically, cocaine is a [1R, 2R, 3S, 5S]-3-(benzoyloxy)-8-methyl-8-azabicyclo-[3.2.1] octane-2-carboxylic acid methyl ester, or the methyl ester of benzoylecgonine. It appears as cocaine base (CAS-50-36-2) and the hydrochloride salt (CAS-53-21-4). There the three components of the molecule, the dotted lines around each component in the drawing mark important parts: ecgonine, methyl alcohol, and benzoic acid (Fig. 30).

Cocaine is still classified as a narcotic by the federal government. It is found in Schedule II of the Controlled Substances Act of 1970, and is subject to all the restrictions placed on opioids also found in Schedule II. Cocaine is at present only approved for topical administration. Its primary use is in ENT surgery, particularly of the nose, pharynx, etc. The esters and derivatives of ecgonine, which are convertible to ecgonine and cocaine, are also controlled according to that Convention. Coca leaf is separately listed in Schedule I and is defined by Article 1, Paragraph 1, as: "The leaf of the coca bush, except a leaf from which all ecgonine, cocaine and any other ecgonine alkaloids have been removed". Cocaine is part of the alkaloids contained in the leaves (folia coca) of the coca bush Erythroxylon coca. It is a white, crystal-like powder, and when in the form of crack, cocaine base usually occurs as small (100–200 mg) lumps ('rocks').

Pharmaceutical Cocaine

A process perfected in the 1880s for extracting and purifying cocaine is known as the "Ecgonine Conversion Process". Rather than isolate just the natural cocaine alkaloid from the coca leaf, this process is designed to isolate only those alkaloids, which contain ecgonine, and to recover from them only the ecgonine portion of their molecules. To make cocaine from ecgonine; methanol and benzoyl chloride are added. Direct synthesis of cocaine was achieved in 1902 by Willstatter, but is a difficult process, and not commercially practical.

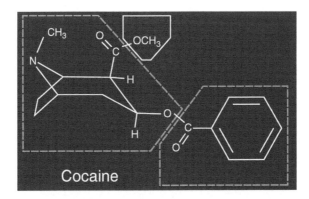

Fig. 30 The molecular structure of cocaine. Only one of its optically active isomers, commonly termed L-cocaine, occurs naturally

Table 6 Summary of the physicochemical properties of cocaine

Appearance	Color- and odorless, bitter taste, crystal like powder
Molecular weight	303.4 g/mol
Chemical formula of the base	$C_{17}H_{21}NO_4$
Melting point	Cocaine hydrochloride 197°C
	Cocaine base 98°C
Solubility of the base	Good solubility in ether
	Not soluble in water
	Good solubility in water (1,800 mg/ml-20°C)
	Solubility of the salt
	Fair solubility in alcohol, chloroform, insoluble in ether
Metabolism	Pseudocholinesterase of blood plasma, liver enzymes (CYP34A)
Mean distribution half time mean time till max. onset of action, nasal	15–20 min
Mean plasma half time mean duration of action, nasal	1–2.5 h
Bioavailability	Oral: 33%
	Nasal: 19%
Excretion	Renal benzoylecgonine and ecgonine methyl ester

Physicochemical Properties of Cocaine

Cocaine has the following physiochemical properties as outlined in Table 6. Its major breakdown product is benzoylecgonine, which appears in greatest quantity in the urine, and consequently this is what it is tested for. Cocaine itself appears in varying quantities from 5% to 20% in its unchanged form in the urine.

Decomposition of Cocaine

Decomposition of cocaine is mainly by means of pseudocholinesterase of the blood and partly by liver enzymes [4] resulting into a number of metabolites (Fig. 31) such as ecgonidine, norecgonidine methyl ester, norecgonine methyl ester, and m-hydroxy-benzoylecgonine [5]. Among these (Fig. 32), benzoylecgonine methy lester is the major metabolite, which is formed both nonenzymatically as well as through the action of esterases found in a number of tissues including hepatocytes, the enzymatic mechanism being the dominant one [6], after which the metabolite is excreted via the kidneys [7, 8].

The simultaneous administration of cocaine and alcohol results in a pharmacological interaction at pharmacodynamic and pharmacokinetic levels. The latter involves an alteration of cocaine kinetics and metabolism, as well as the biosynthesis of newly active metabolites, such as cocaethylene [9]. Cocaethylene is metabolized along the same pathways as cocaine (Fig. 33). Its detection in biological samples indicates the combined consumption of cocaine and alcohol. During the interaction of both substances, the rise in cocaine plasma concentrations can explain many of cardiovascular [10, 11] and behavioral effects observed [12].

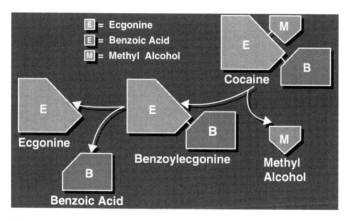

Fig. 31 The primary pathways in the decomposition of cocaine to its main metabolite benzoylecgonine, which can later be traced in an urine analysis

Cocaine, as a Local Anesthetic: Mechanism of Action

Cocaine's use as a local anesthetic had been promoted by Koller since 1884. The mode of action is that in the resting state, an equilibrium is reached when the axon interior

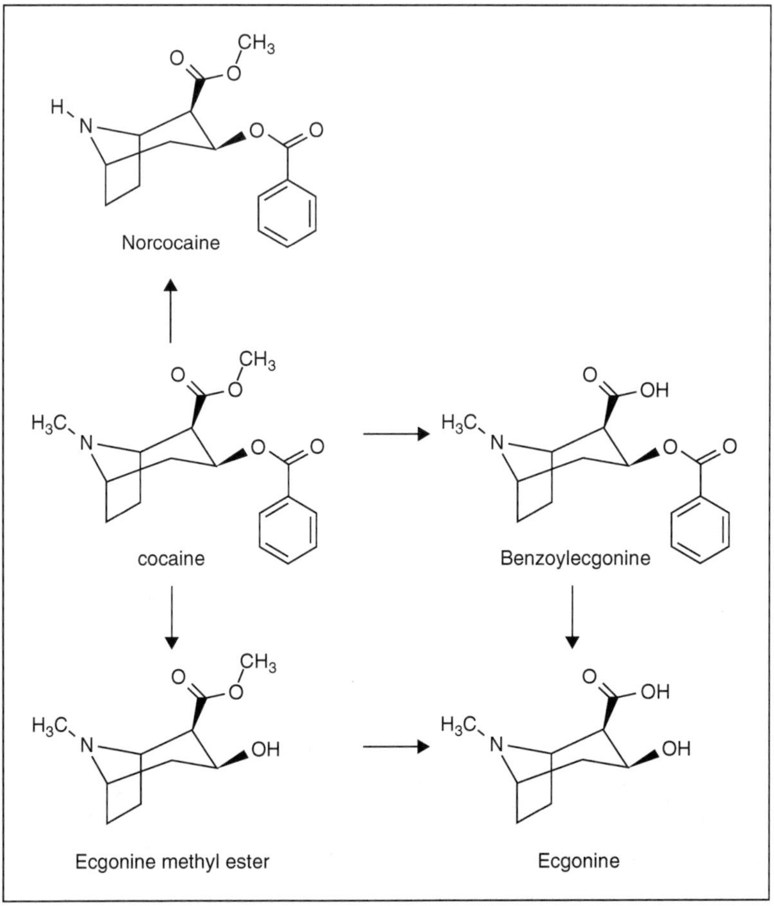

Fig. 32 Molecular structure of the different metabolites of cocaine being formed by pseudo-cholinesterase and liver enzymes

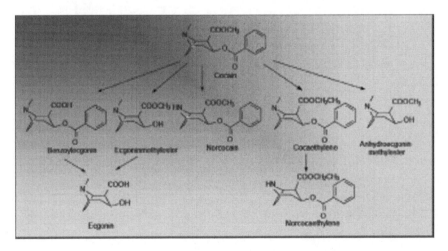

Fig. 33 Formation of cocaethylene when cocaine is ingested together with alcohol

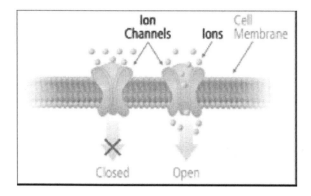

Fig. 34 Cocaine prevents migration of sodium ions into the axon of the nerve cell during stimulation of the nerve. This prevents the sodium ions from changing the electrical charge across the membrane and thus prevents the nerve from an action potential

is negatively charged with respect to the exterior. When the nerve is stimulated, activation gates in the sodium channels of the membrane begin to open, allowing sodium ions to migrate into the axoplasm, or inside of the cell. This causes the voltage difference to be reduced between the inside and outside of the axon. As sodium channels are then closed by inactivation gates, other ions move out of the axon to repolarize the membrane. Cocaine blocks the sodium channel preventing sodium ions from passing into the axoplasm (Fig. 34). This effectively prevents the sodium ions from causing a reduction of the voltage difference between the inside and outside of the axon, and this inhibits conduction of the nerve impulse.

Cocaine is available today from several manufacturers (DuPont, Merck Pharmaceuticals, Roxane, Mallinkrodt). Crystals are available from Mallinkrodt for preparing topical solutions. Most commonly, however, cocaine is packaged as premixed solutions of 4% and 10% containing 40 mg/ml and 100 mg/ml respectively, for Topical Solutions (Roxane; Fig. 35). These solutions are strictly for topical use only. However, they have been known to be used orally as adjuncts to opioids in the management of severe pain.

Cocaine is the only local anesthetic agent with vasoconstricting properties. This makes it especially useful to ENT procedures. Other (synthetic) local anesthetics usually must be combined with a vasoconstrictor such as epinephrine to keep their action localized to the site of surgery. While cocaine gained fame early as an anesthetic for eye surgery, it was later discovered that it causes clouding and pitting of the cornea and, in some cases, ulceration as well. It is because of these side-effects that topical cocaine solutions are not for ophthalmic use! Due to this, and the concern over systemic toxicity, cocaine has largely been replaced by synthetic agents such as lidocaine, bupivacaine, levobupivacaine, ropivacaine, tetracaine and/or procaine, all of which show a similar basic structure of an amino group linked via a carbon atom with an aromatic group (Fig. 36).

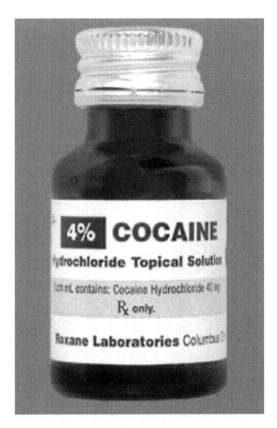

Fig. 35 Although being replaced by synthetics and being a scheduled II drug, use of a premix solution of cocaine is still available, strictly for topical anesthesia as a topical agent in ENT surgery

Cocaine's CNS Effects: Mechanism of Action

Of perhaps more interest than the local anesthetic effects of cocaine are the CNS effects, for it is this area that is responsible for the widespread popularity of cocaine as a substance of abuse (Fig. 37). Being also acclaimed as a cure for morphine and other dependencies, Freud was one of the first to experiment with cocaine on his friend Ernst von Fleischl-Marxow (Fleischl), morphine dependence, who suffering from neuropathic pain from an amputated thumb. At the same time countless patent medicines and tonics containing cocaine were available.

The central nervous effects of cocaine are related to the neurotransmitters norepinephrine and dopamine, which are normally released from presynaptic vesicles, diffuse across the synaptic cleft and bind with receptors on a receiving neuron. Excess norepinephrine (or dopamine) is then re-uptaken into the presynaptic vesicles for later use. Cocaine has the ability to block the re-uptake mechanism, preventing the re-absorption of the neurotransmitter. This results in higher

Pharmaceutical Cocaine

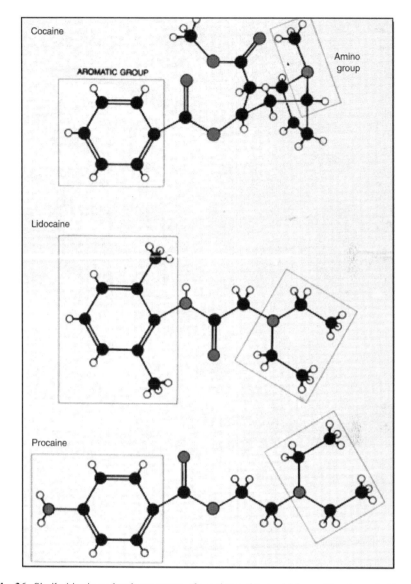

Fig. 36 Similarities in molecular structure of cocaine and two widely used local anesthetics

concentrations of the neurotransmitters norepinephrine and dopamine remaining in the synaptic cleft, which in turn bind to more post-synaptic receptor sites (Fig. 37).

Such an excess of neurotransmitters results in a concomitant overload at the receiving neuron, which then fires more frequently.

Primarily, euphoria is an effect of cocaine use and it is the one that gets all the attention anyway. There are however, some other important effects: increase in awareness; probably due to increased catecholamine activity. This along with euphoria is probably caused by alteration of dopaminergic activity in the limbic system. Tachycardia,

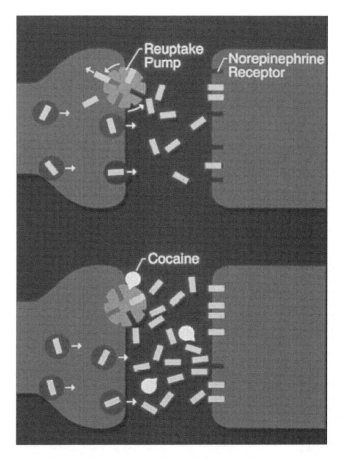

Fig. 37 Cocaine acting presynaptically as a reuptake inhibitor at the dopaminergic, serotinergic, and noradrenergic transmitter system

increased blood pressure and respirations are likely due to stimulation of norepinephrine, resulting in an "overamped" CNS (Fig. 38; Table 7). Anorexia seems to go hand-in-hand with CNS stimulants, and is one of the effects seen with chewers of the coca leaf, where coca was used to "deaden man's hunger and thirst".

Table 7 Summary of effects of cocaine on the sympathomimetic system

CNS effects of cocaine
Euphoria
Sensory awareness is increased
Higher state of vigilance
Tachycardia
Blood pressure rises
Anorexia
Respiratory increase
Mind widening
Increase in creativity
Reduced need for sleep

Pharmaceutical Cocaine

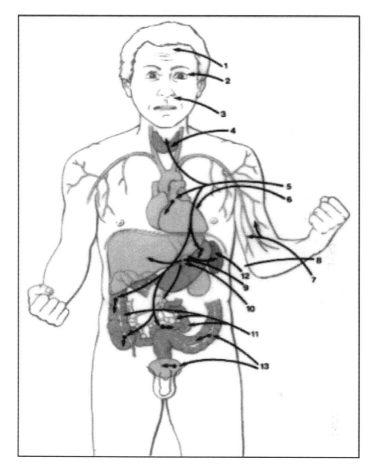

Fig. 38 Consequences by cocaine ingestion resulting in a "fight or flight reaction" with CNS activation (effects 1–12 see below)

In summary, cocaine will induce the following effects

1. Increase in vigilance with activation of defense mechanisms from the limbic system, reduced need for sleep
2. Mydriasis
3. Nasal drip
4. Release of thyroid hormone followed by glucose release
5. Coronary vessel constriction
6. Supraventricular tachycardia, increase in myocardial contractility, increase in blood pressure
7. Vasodilatation with increased muscle perfusion
8. Vasoconstriction of blood vessels of skin

9. Release of epinephrine and norepinephrine
10. Mobilization of glycogen from muscle and liver
11. Inhibition of motility and reduced perfusion of stomach and small intestine and inhibition of digestive function
12. Increase in muscle tone of the large intestine

Concerning the autonomic nervous system, Erlanger and Gasser stated: "Cocaine enhances the response of the innervated structures of the sympathetic system, as well as epinephrine and norepinephrine, both sympathetic chemical transmitters. In this way, they provoke vessel constriction, mydriasis and tachycardia being related as sympatho-mimetic actions. It is for this reason, that medium cocaine doses produce tachycardia and an increase of blood pressure on the heart. And, as in every excess, high doses revert their action and blood pressure can decrease intensely". This mechanism of action of cocaine as a stimulant is based on the blockade of recapturing norepinephrine in the synaptic cleft. Once the mission of the neurotransmitter norepinephrine, which is in charge of carrying the message via the synaptic cleft to the other cell, has been accomplished, it is degraded by special enzymes or recaptured. The cocaine molecule blocks the recapturing of the neurotransmitter, forcing it to remain in the synaptic cleft for a longer period of time and continue its stimulating action (Fig. 37). When used over a long period of time, another clinical effect is related to blockade of re-uptake of neurotransmitters. Such long-term prevention in re-uptake can cause depletion of the neuron's supply (Fig. 39), and it is this phenomenon that is theorized to account for the depression and craving associated with the cessation of cocaine use.

Emotional Effects of Cocaine

Since Sigmund Freud popularized the uptake of cocaine crystals prepared by the Merck Laboratories (Fig. 18), through the nasal mucous membrane, other possible ways of administration were set aside for reasons not too well determined. This was in spite the fact that he himself, due to injuries in his nasal passages, resorted to oral absorption dissolving the powder in water to drink. Actually, the application of the powder in any of the mucous membranes (nasal, oral, genital, anal) has the same speed of absorption. However, nasal aspiration continues to be the preferred method for easier administration. A quantity of 60 mg dispensed through the nasal mucosa, after entering the brain circulatory system, produces a general increase of body energy. Metabolism accelerates in proportion to the dose, just as the physical, intellectual activities and the emotional tone increase. Cardiac frequency and blood pressure also rise slightly. Respiratory frequency and volume are also increased thus requiring a greater oxygen intake. Aside from such acceleration of physiological functions, hunger and tiredness sensations disappear. There is an increase of intestinal peristalsis, which, however, is not observed in coca chewing, and results in lesser assimilation of nutrients. The latter, in association with the increase of metabolism, individual

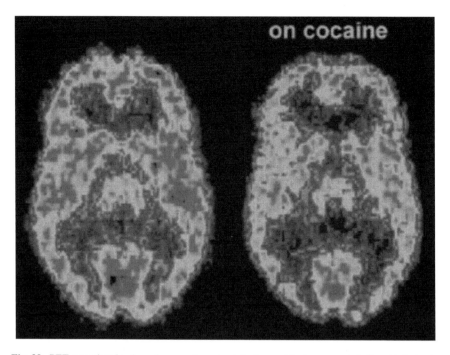

Fig. 39 PET scan showing how the use of cocaine interferes with glucose-metabolism in the brain. The dark color denotes the highest level of glucose utilization as seen for the normal brain on the left. Yellow represents less utilization and blue the least utilization. Glucose provides energy to neurons and when those neurons cannot use glucose, many brain functions will be disrupted

activity and the anorexia effect, explains the success achieved in weight reducing treatments. In addition to snorting and intravenous use, sublingual application has also been reported [13], where street cocaine (cocaine hydrochloride) is mixed with baking soda and placed under the tongue. Since the more neutral the pH of that mixture the higher absorption through the capillary bed under the tongue resulting in better titration to effect with lesser transitions between highs and lows (crashes), less interference with eating and sleeping and lastly, no nasal drip as usually experienced after snorting.

Aside from physiological functions, emotional areas are also stimulated (Fig. 39). There is a pleasant peaceful feeling, and at higher doses one feels euphoria and enthusiasm. The user feels content, loquacious (talkative), uninhibited and courageous. If there were emotional depressions before, it subsides rapidly, giving way to positive feelings about life in general. In regard to the psychological effects, one can observe an acceleration of all higher mental functions, such as capacity for thought course and association of ideas. The individual is wide-awake, and sleep can be postponed. The capacity for abstraction, concentration, attention and memory improves. The content of thoughts is enriched and fantasizing or having more optimistic and pleasant ideas about life are facilitated. Sensorial perception becomes keen and sensitive. In respect to sexual activity, cocaine has very definite aspects. It increases sexual desire and potency notoriously, however, chronic cocaine users show a significant decline in sexual behavior.

The two-sided and sometimes schizophrenic opinion of the society regarding the use of cocaine is reflected in the media. While there are articles describing the devastating effect of long-term use and abuse, in the late 1970s the use of cocaine by championship athletes was associated with manliness and power. Cocaine became the media's powdery star with stories of its use by the famous, its increasing expense and its description as the caviar or champagne of drugs. In July 1981 Time magazine showed a champagne glass full of sparkling white powder with the headline "HIGH ON COCAINE- a drug with status- and menace". The net effect of the article was a positive and alluring image of cocaine. In New York magazine (1978) cocaine was referred to as "the drug of choice" and "every bit as enjoyable as Freud claimed": "Cocaine is God's way of telling you that you've made too much money."

Addictive Properties of Cocaine: Mode of Action

Although not fully understood as yet, a great deal of evidence suggests that the action of cocaine at central dopaminergic (and other adrenergic) neurons may provide the underlying cause of the craving for cocaine. The blocking of the re-uptake of dopamine and subsequent enhanced activity of this neurotransmitter at dopaminergic receptor sites located in the limbic system of the brain, may account for the euphoria associated with cocaine use. The repeated blocking of dopamine in chronic users may lead to a state of functional deficiency of dopamine in these individuals. By blocking re-uptake of the neurotransmitter, less is available in the presynaptic vesicles for use in normal neuronal regulation. In the case of dopamine, that may lead to depression in the user. What is especially insidious is that this depression might last for days or weeks after the last use of cocaine, forming the basis for craving (i.e. drug hunger), which ultimately will end up in bodily harm (Fig. 40).

It is this craving that is thought to be the cause of psychological dependence seen with cocaine use. Taking this one step further, the behavioral pattern known as *addiction* may be fostered by such craving. Thus, the biochemical (physical) cause can manifest itself in a psychological symptom (craving), and ultimately form the basis for a behavioral pattern known as *addiction*.

According to Jerome Jaffe in Goodman & Gilman's [15] the definition of "*addiction*" consists of various elements. Each element describes a behavioral pattern, devoid of any mention, of physical effects, tolerance, or underlying psychological etiology. It matters not why the individual has the condition and engages in this pattern of behavior, but what the essential elements of that behavior are.

(a) One notifies first: Compulsive use of a substance, possibly related to physical and/or psychological effects as noted above.
(b) Second: Securing a supply of the substance (i.e. the hunt) the preoccupation with getting it at any cost, disregarding physical, financial, legal margins.
(c) And finally: A high tendency to relapse after withdrawal of the substance.

It is believed that repetitive overactivation of the dopaminergic reward system within the deep brain structures (Fig. 41) is the cause for an addictive behavior pattern. If physical dependence was all there was to "addiction", withdrawal could cure the problem. Once there is no more drug in the system, no more "addiction" would result.

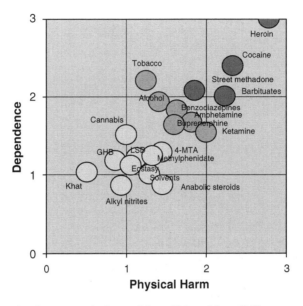

Fig. 40 A rational scale to assess the harm of drugs (Adapted from [14])

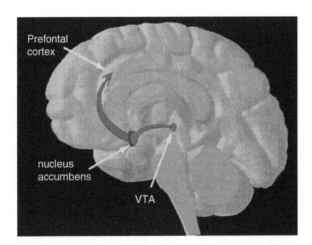

Fig. 41 The dopaminergic reward system in the deep brain structures (mesolimbic system) with the VTA (= ventral tegmental area), where mostly all addictive agents exert their action. It is here where cocaine directly activates the nucl. accumbens resulting in an increased release of dopamine with ensuing euphoria (Modified from [16, 17])

However, there is more to "addiction" than the mere presence of the drug itself. It is this fact more than any other that is persuasive for labelling cocaine as addictive. Presence of the drug on a chronic basis is not necessary. Therefore users who use only once a week, for example, may indeed fit the description of **"addiction"** if' they feel compelled to use, they are driven by craving the "drug hunger", they are preoccupied with the "hunt" for the drug, taking economic, social and legal risks in the process, and continue to use it even after expressing a desire to stop resulting in a relapse. It is said that "addiction" occurs when the user doesn't feel "normal" without.

> **Addiction** "A behavioral pattern of drug use, characterized by compulsive use, the securing of its supply, and a high" tendency to relapse after "withdrawal". Adapted from [18]

Contrary to addiction, the label ***drug abuse*** alone is insufficient to describe every use of illegal drugs, even cocaine. In order to get a perspective on what constitutes abuse one needs a clearer picture of what psychoactive substances can do to mankind. Although ***physical dependence*** to some is the same as "addiction" there are subtle differences. Being exposed to the concept of physical dependence through movies and television, where addicts go into withdrawal if they can't get a "fix" (opioid users), the concept of physical dependence became synonymous with "addiction". It is this simple mistake that has led many to dismiss cocaine as *non-addictive* or *non-dependence producing* substance. A quick review of many medical textbooks will confirm that there is a long-held belief that cocaine does not produce **physical dependence** because there isn't a definitive withdrawal syndrome associated with it, nor is there tolerance to its effects. This erroneous belief was based on trying to compare the effects of cocaine to those of narcotics (i.e. opioids). As will be seen, when cocaine is looked at alone, it indeed has a unique withdrawal syndrome, and fits the definition of physical dependence set forth here. But physical dependence alone is not ***addiction***.

> **Physical Dependence** It is an altered physiological state produced by repeated administration of a drug, necessitating its continued administration to prevent the appearance of a withdrawal or abstinence syndrome.

Psychological Dependence

It is here that things really get complex, as we are now dealing with concepts that are not as verifiable with objective tests such as blood tests, or X-rays. This is not to suggest that there isn't a biochemical basis for the effects described here but that these symptoms are usually manifested in psychological or behavioral abnormalities that are a product of several forces, environmental, biological and social. The feelings range in intensity, and can be mild or overwhelming in nature. It is essential to understand that the craving ("drug hunger") is not necessarily related to any physical need. That is, the user does not have to be physically dependent in order to crave the substance. Acceptance of this fact will help in divorcing "addiction" from simplistic explanations such as mere physical need. When **physical dependence** is also a factor however, it can become a powerful secondary reinforcer of craving, so that now the user not only has hunger for the substance, but is also pushed into further use by the prospect of unpleasant side effects. A more complicated explanation is related to the positive vs. negative reinforcing effects. To summarize, it is necessary to separate the various forces that act upon a user and lead to the behavioral condition defined as psychological dependence.

> **Psychological Dependence** Intensity may range from mild desire to strong compulsion.
> Craving is not necessarily related to any physical need. However, when physical dependence becomes established after chronic use, it becomes a powerful secondary reinforcer of psychological dependence.

Cocaine is highly addictive – once a person has used cocaine the urge to use it again is almost irresistible, and in this respect it is one of the most insidious "recreational" drugs around. However, despite the feelings of overpowering a drug like ecstasy brings, cocaine never produces satisfaction. The user is left with an irresistible craving to experience the joy of cocaine over and over again, which leads to tolerance and an ever-increasing dose. Laboratory research has shown that given an option, animals prefer cocaine to food, water and even sex. If given free access to it, they continue to take the drug until they overdose and die. The same, unfortunately, is true for humans as well. For this reason cocaine is classified as an illegal substance in most countries of the world.

Cocaine Intoxication: Strategy for Treatment

The vast majority of patients, who are alive when emergency personnel get to them, will survive cocaine poisoning. If someone is going to die, they will usually do so within 2–3 min of first experiencing symptoms of advanced CNS stimulation. Unless this occurs in the controlled medical environment of an emergency room (ER) or a fully equipped ambulance, death is quite likely. Reporting in the Annals of Emergency Medicine in October 1982, George (Skip) Gay of the Haight-Ashbury Free Clinic of San Francisco, set down what has become standard in the description of cocaine toxicity [19]. His so-called "Caine", or "Casey Jones" reaction can be divided into three "phases" or stages of signs:

Caine or Casey Jones Reaction

Early Stimulation Phase		
CNS	*Circulation*	*Respiration*
Excitement	Brady/Tachycardia	High respiratory rate
Euphoria	PVC`s	High tidal volume
Soaring	Blood pressure rise	
Garrulous		
Muscle twitching		
Teeth grinding		
Nausea		
Headache		
Preconvulsive movements		
Body temperature increase		
Advanced Stimulation Phase		
CNS	*Circulation*	*Respiration*
Tonic-clonic seizure	Hypertonia	Cyanosis
Hyperkinesia	Irregular pulse	Dyspnea
Hyperthermia	High cardiac output	Irregular rate
Status epilepticus		
Depressive Phase		
CNS	*Circulation*	*Respiration*
Flaccid paralysis	Ventricular fibrillation	Respiratory arrest
Coma	Circulatory failure	Cyanosis
Fixed pupils dilated	Cardiac arrest	Pulmonary edema
Loss of vital function	Death	Death
Death		

Phase of Early Stimulation

The effects in this phase are commonly experienced by users, who have at least some, if not all at one time or another, of these reactions to even small amounts of cocaine. Patients are rarely seen in this stage, as they will usually not feel distressed until crossing into phase II (advanced stimulation), when medical attention may be sought. Note that initially there is a bradycardia associated with cocaine use, but by the time medical personnel are asked to evaluate the patient, this has been long gone.

Phase of Advanced Stimulation

It is in this phase that the user begins to feel that something has gone wrong. This may have been signalled at the end of phase I by preconvulsive movements, heart palpitations, headache, nausea/vomiting etc. Now, they begin to experience seizures, their temperature rises precipitously, cardiovascular effects are much more pronounced, breathing may become impaired, and often there is a sense of impending doom ("I think I'm gonna die"). At this point it is worth noting that the progression through the phases of stimulation can be very rapid, and accelerate at any time from one phase to the next. A patient who seems quite stable 1 min, can progress to severe depression of the CNS in the next.

The Depressive Phase

By this time it's usually all over except where to send the body. There is a sudden loss of control, the consequences of an "overamped CNS". In the first two phases, we see overstimulation, and in the third and final phase, there is understimulation. The disruption of the CNS by "relentless stimulation" leads to the depression of its function in maintaining circulatory and respiratory activity. Patients die quickly, due to an electrical-imbalance of the cardiovascular system. The place to intervene then, is before they pass out of phase II.

Basics in Treatment of Cocaine Overdose (OD)

Patients presenting an altered mental status should be treated with:
1. ABCs (Airway, Breathing, Cardiovascular status)
2. Administration of dextrose (in case of potential hypoglycemia)
3. Administration of thiamine (for preventing Kosakoff's syndrome in potential alcohol intoxication), and
4. Administration of 2 mg of naloxone (Narcane®) for reversal of porential opioid overdose

All this may effectively treat or rule out a number of other potential causes or contributors to the patient's condition, including hypoglycemia and opioid poisoning. It must be remembered that as many as a quarter of patients who present with cocaine intoxication also have opioids on board. Since other drugs may be interfering with the diagnosis at this point, and until toxicological studies are performed, no definitive cause can be established. For this reason, a high index of suspicion for the presence of depressants such as alcohol and opioids must be maintained, with appropriate measures (naloxone etc.) ready to deal with a sudden onset of these effects.

Advanced Treatment of Cocaine OD

The overdose must realistically be treated as a *stimulant* overdose in the most general sense, since until a positive toxicological screen is obtained, one cannot be certain as to the offending agent. Often even the patient is unaware of the actual composition of the substance they ingested (see chapter on adulterants in cocaine). Certain "active" cuts such as amphetamine, PCP, phenylpropanolamine, and/or yohimbine can last longer and be more cardiotoxic than the cocaine alone. Treatment is therefore directed at target symptoms, with pharmacologic therapy designed to stabilize the patient until the effects of the poison have abated. Thereafter intervention continues for cardiac symptomatology, with the goal to quiet the irritable heart. Propranolol was the most frequently mentioned beta-blocker to accomplish this task, but it was by no means the only effective agent [20]. However, due to its alpha-adrenergic action it will result in peripheral vasoconstriction, thus further increasing the elevated blood pressure. Therefore the β-blocker of choice is esmolol (Brevibloc®, Baxter), which is a pure β-blocking agent, with a very short duration of action (ca 7 min). By this last property the dose essentially can be titrated to effect. There has been some discussion of the utility of calcium channel blockers (verapamil and others) but all the recipes are based on clinical judgement rather than controlled studies. So in reality it is rather a trial-and-error therapy. The goal of beta-blocker therapy is to lower the diastolic pressure to about 90 mmHg and stabilize the arrhythmia.

Selective Approach to Cocaine OD

The most serious and life-threatening reactions to cocaine poisoning are seizures. Often the patient seizes soon after a dose of cocaine, usually without warning. This can be followed by an "emergence arrhythmia" which can be manifested as ventricular fibrillation from which there is little chance of recovery unless it occurs in a controlled medical setting such as an ER. Therapy is commonly intravenous diazepam or a short-acting barbiturate. Haloperidol (Haldol®) has been mentioned in the literature, but it should be kept in mind that this agent can lower the seizure threshold. Before considering pharmacologic therapy, be certain that the patient's temperature is stable. As Skip Gay MD admonishes: *One has to cool these people down as they are cooking their brain cells, your drugs won't do any good if you can't get control of hyperthermia*. An ice-bath is recommended, and to be continued to below 39°C or 102°F until the patient's heart rate is down [19].

Avoiding Procedures in Cocaine OD

One of the commonest mistakes that can be made in treating a patient with possible cocaine overdosage is to resort to "P3" therapy. While there is a great temptation to restrain, drug and pack these people off to the loony bin, this approach can and does contribute to the deaths of patients who might otherwise recover. With the unpredictable nature of cocaine (and other stimulants), a rapid acceleration through the phases of the "Caine reaction" can occur. Unless the patient is kept in a controlled medical (ER) environment, where the goal is to treat target symptoms and avoid further stress to the individual's CNS, the outcome may be fatal.

If confronted with an individual with a toxic delirium, avoid the "P3" therapy, which is

1. **P**hysical restraint
2. **P**henothiazines in staggering doses
3. **P**sychiatric ward where the user is being locked up and left on their own

In contrast, one should use the "The Science of **ART**" named after the Haight-Ashbury Free Clinic. It is based on the theory that a patient with an "overamped" CNS needs calming more than anything else. When dealing with an individual who has taken what has been called by some *the premier ego enhancing drug known to man* and who may have dealt on the *Faustian point* (i.e. sold their souls for cocaine), The first consideration should be

1. **A**cceptance of their state. Acceptance only means a sincere approach to explaining to the patient that they've gone too far in the pursuit of pleasure, and now it's time to come down. The immediate goal is to
2. **R**educe stimuli (dark area of the ER if practical), and give rest and assurance that help is the only desire of the medical staff. And finally, a good
3. ***T**alkdown* technique will do more than many drugs to calm the CNS

Special Pathologies in Chronic Cocaine Use

The Toxic Cocaine Delirium

Cocaine produces an array of neuropsychiatric states, and the violently agitated patient is perhaps the most dramatic ED presentation caused by this drug. This person typically arrives with a multitude of personnel who are attempting physical control. There is an appearance of incredible strength, but this may reflect an indifference to pain; the risk of physical harm to both caregivers and the patient is significant. The patient is delusional, paranoid, and often overtly psychotic. This state does not clearly relate to cocaine dosage, blood level [21], route of administration, or frequency of use. It does not seem to be precipitated by adulterants or concomitant use of other drugs, with the possible exception of alcohol; the role of cocaethylene.

While there is an increasing rate of use among young adults in all western countries over the past years (Fig. 42), more often, the combination of cocaine with alcohol is being advertised as this allegedly results in a prolongation of effects. Some 99% of contemporary Western users mix cocaine and ethyl alcohol. Cocaine and alcohol combine to be metabolized to another hugely reinforcing compound, cocaethylene. Coca-use only really took off in the West when it was blended with an alcoholic beverage. However, such combination is substantially more toxic than either substance on their own. For instance, about twelve million Americans combine ethanol with cocaine to produce a pronounced and prolonged euphoria. However it is known that the combination of these two drugs is substantially toxic. This may be a result of cocaethylene, detected in the serum of emergency patients. Cocaethylene is a cocaine metabolite, formed in the liver only in the presence of ethanol [9]. Being pharmacologically active it depresses the myocardium causing cardiotoxicity [22].

Several studies have demonstrated that cocaethylene's effects are likely additive to cocaine resulting from the direct action of each drug [23, 24]. Thus, the combination of alcohol and cocaine seems to produce more abnormal locomotor behavior than any drug alone [25]. Untreated, this confusional, agitated state may last for several hours, far longer than cocaine's other effects. It often is followed by a period of complete exhaustion and obtundation.

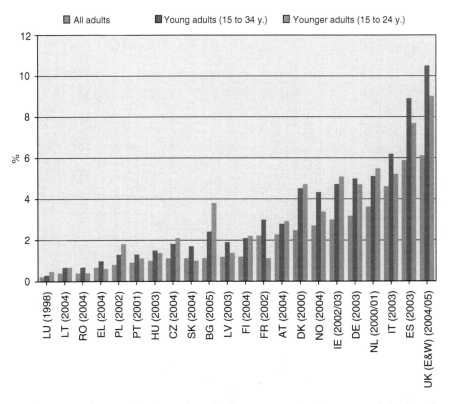

Fig. 42 Data on the recreational use of cocaine from recent national surveys available in each country (From Epidemiological tables on population surveys, Statistical Bulletin 2006)

Cocaine and Myocardial Infarction

The pathophysiology of cocaine-induced myocardial ischemia is multifactorial [26, 27]. Proposed mechanisms are coronary thrombosis, coronary artery vasoconstriction, mismatch between myocardial oxygen demand and supply, and accelerated atherosclerosis [28, 29]. The extent to which the mechanisms may interact is unknown. Coronary thrombosis can develop in the presence of normal or diseased coronary arteries, possibly as a result of alterations in platelet and endothelial functions. Studies have proven that cocaine increases human platelet activation and aggregation [30]. Additionally, vascular spasm may cause damage to the endothelium, creating a nidus for platelet aggregation and fibrin deposition and resulting in thrombus formation [28, 31].

Coronary artery vasoconstriction or spasm results from alpha-adrenergic stimulation and may occur in patients without coronary artery disease. Although this mechanism is not completely understood, it is known that it differs from Prinzmetal's angina [28, 31]. In addition, cocaine has sympathomimetic effects that induce tachycardia and hypertension, resulting in an increased myocardial-oxygen demand. When this demand exceeds the supply, myocardial ischemia occurs. These sympathomimetic effects most likely act synergistically with other mechanisms to cause ischemia and are exacerbated by concomitant cigarette smoking [28, 31].

Chronic use of cocaine may lead to premature atherosclerosis. In autopsies, coronary atherosclerosis has been found with increased prevalence in young cocaine users compared with age-matched non-using controls [28, 30, 31].

The differential diagnosis of cocaine-associated chest pain is similar to that of chest pain unrelated to cocaine use, but may vary depending on the route by which the drug was ingested. The smoking of crack cocaine has been associated with alveolar rupture resulting in pneumothorax, pneumopericardium, and pneumomediastinum. An ailment known as "crack lung syndrome," which involves pulmonary hemorrhage, chest pain, pulmonary edema, and an interstitial lung process, can occur. In addition, asthma, pneumonia, and pulmonary vascular disease must be considered. In patients using IV cocaine, endocarditis should be ruled out, especially if the patient also has fever. Finally, although it is a rare condition, aortic dissection must be considered in any patient with chest pain and a history of cocaine use because of the high mortality rate of roughly 27% [28, 29]. ECG interpretation in patients with cocaine-induced chest pain is problematic. Approximately 60% of patients with cocaine-induced MI have a nondiagnostic ECG, and 56–84% of patients with cocaine-associated chest pain have abnormalities present on ECG [32]. Thus, in cocaine users, MI cannot be ruled out in the setting of a normal initial ECG, nor can it be concluded that a patient needs reperfusion therapy if abnormalities are present on ECG [28, 29]. The most useful diagnostic tool for detecting cardiac injury in this patient population is serum biochemical markers. Cardiac troponin I and T are the most specific, and are preferred. Elevations in creatine kinase (CK) and CK-MB occur in the absence of myocardial ischemia due to cocaine-induced skeletal-muscle injury. It has been reported that, after using cocaine, approximately half of patients have elevated serum CK with or without myocardial injury [29, 33]. Urine drug testing also may be useful, especially in patients who initially deny using cocaine. The metabolite benzoylecgonine can be detected for up to 48–72 h after cocaine use [29] (see also chapter on urine drug testing).

Treatment of Cocaine-Related Myocardial Ischemia (MI)

The treatment of patients with cocaine-related ischemia or MI varies only slightly from the traditional treatment of acute coronary syndrome (ACS). All patients should be administered oxygen and placed on a cardiac monitor. Based on extensive investigation in patients with ischemic heart disease unrelated to cocaine, a favorable safety profile, and theoretical considerations, aspirin should be given to prevent the formation or extension of thrombi if there are no contraindications (i.e., allergy or suspected subarachnoid hemorrhage) [29, 33]. Initial therapy also should include nitroglycerin. Nitroglycerin has been shown to reduce infarct-related complications and limit the extent of acute MI in patients with ischemia unrelated to cocaine. Studies also indicate that nitroglycerin alleviates cocaine-induced vasoconstriction and relieves symptomatic chest pain [34–36].

Benzodiazepines, in particular lorazepam, also have established benefits. Early use may decrease cocaine's cardiovascular toxicity by decreasing its central stimulatory effects [37]. The combination of lorazepam plus nitroglycerin appears to be more effective than either agent alone for relieving chest pain associated with cocaine use [38]. The same was not proven for diazepam in a similar study, however [39].

Although used in the treatment of coronary ischemia that is not related to cocaine use, beta-blockers are contraindicated for cocaine-associated ischemia. Presumably through unopposed alpha-adrenergic stimulation, beta-blockers enhance coronary vasoconstriction and increase blood pressure [28, 40]. They also increase the likelihood of seizures and may decrease survival [37]. Although labetalol has been used safely in some patients, it is not recommended based on controlled studies performed in animals and humans. Labetalol has combined alpha-beta antagonism, but the beta antagonism is far more potent [29, 37]. Owing to conflicting data, the role of calcium-channel blockers in the treatment of cocaine-associated ischemia has not been established. Cardiac-catheterization studies in patients with cocaine-induced coronary vasoconstriction found that verapamil reverses the vasoconstrictive effects of cocaine, but large-scale clinical trials have shown no benefit in acute MI unrelated to cocaine use [29]. The American College of Cardiology/American Heart Association's 2007 guidelines for the management of patients with unstable angina/non-ST-elevation MI, however, recommend calcium-channel blockers in combination with nitroglycerin for patients with chest pain after cocaine use [34].

Phentolamine, an alpha-adrenergic antagonist, also can be used to achieve vasodilation in patients who continue to have chest pain after administration of oxygen, aspirin, benzodiazepines, and nitroglycerin [41]. One case report describes a 38-year-old man with cocaine-associated chest pain refractory to oxygen, diazepam, and nitroglycerin that resolved after low-dose phentolamine [42]. To avoid hypotension while maintaining the anti-ischemic effects, phentolamine 1 mg was recommended for such patients [42].

Thrombolytic therapy should be considered in patients having ST-segment elevation MI only when cardiac catheterization is impossible. Although cocaine's known thrombogenic properties make thrombolytic therapy attractive in theory, adverse outcomes have been documented in several case reports. When it is balanced against the low mortality seen in patients with cocaine-associated MI, thrombolytic therapy's risks most likely outweigh its benefits in this patient population [29].

Secondary Prevention of Cocaine-Related MI

The key to secondary prevention of cocaine-related chest pain is the cessation of cocaine use. Sadly, 60% of patients use cocaine again during the year following an episode of chest pain [43]. Cocaine-related death, MI, and recurrent chest pain are extremely rare in patients who stop using cocaine. Tobacco should be avoided as well, as it is a major contributor to the risk for coronary artery disease; it also is associated with a faster onset of chest pain and vasoconstriction after cocaine use [44]. Patients likely will benefit from conversion of other traditional risk factors for heart disease such as high cholesterol, high blood pressure, and obesity. The use of aspirin to prevent platelet aggregation also may be beneficial for secondary prevention. The role of calcium-channel blockers and nitrates remains unproven, and beta-blockers should be avoided in any patient who may use cocaine again [43, 44].

Cocaine-Related Rhabdomyolysis

In addition, the presence of rhabdomyolysis should be considered in cocaine-intoxicated patients. In one study [45] 24% of all patients presenting a cocaine-related complaint were found to have rhabdomyolysis, defined as a creatinine kinase (CK) level at least five times normal. Most of these cases were mild and asymptomatic. Myoglobinuria is another indicator of muscle necrosis, and patients may be rapidly screened for this with a urine dipstick for occult blood (detects hemoglobin or myoglobin); positive results necessitate assays for CK, renal function, and electrolyte disturbances.

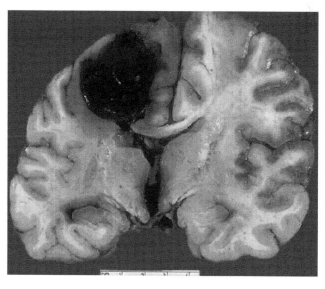

Fig. 43 Representative post mortem section with intracranial bleeding following free base smoking in a subject

Cocaine and Cerebrovascular Ischemia/Hemorrhage

Other potential neurologic catastrophes include subarachnoid hemorrhage, intracerebral bleeds, and ischemic strokes. Cocaine produces an intense, short-lived burst of systemic hypertension that may lead to hemorrhage (Fig. 43).

Subarachnoid hemorrhage (SAH) due to cocaine presents, as do nondrug-related SAHs, with severe headache, loss of consciousness, nausea/vomiting, and/or neck stiffness. Of 27 reported cases reviewed by Green and coworkers [46], 78% were found to have an underlying predisposing condition such as an aneurysm or arteriovenous malformation. Other forms of intracranial hemorrhage due to cocaine are present with headache, seizure, altered sensorium, and/or lateralizing findings. Approximately half the reported cases of intracerebral hemorrhage were found to have an underlying vascular lesion [46]. Cocaine may induce a bleed due to its transient hypertensive effect as in SAH or due to hemorrhage at a site of ischemia produced by vasospasm. Ischemic cerebrovascular accidents may occur as often as hemorrhage in free base smokers [47], although both are fortunately rare events. Cocaine-induced hypertension and vasospasm appear to be the most likely mechanisms of ischemic stroke [48]; thrombosis and embolic phenomena often seen with arrhythmias and cardiomyopathies are also possible [49].

Summary of Advanced Treatment in Cocaine OD

> **Since the cause of an altered mental status in a person before having a lab test is never clear, initial treatment should comprise of the following**

1. ABC's (airway, breathing, cardiovascular support)
2. 100 ml 50% dextrose intravenously for reversal of a potential hypoglycemia
3. Thiamine 100 mg intramuscular for a potential alcoholic intoxication and for prevention of a Korsakoff syndrome
4. 2 mg naloxone intravenously for reversal of a potential opioid overdose

This is followed by a particular regimen treating the symptoms because a specific antidote is not available. Special attention should focus on potential cardiac arrhythmias as they are the prodromi of detrimental ventricular fibrillation (Fig. 44).

Treatment of Tachyarrhythmia

- Labetolol (0.25 mg/kg), or
- Esmolol (100 mg/kg/min), or
- Lidocaine (50–100 mg), or a second-line
- Ca^{++}-channel blocker (i.e. Verapamil 2.5–5.0 mg)

Treatment of Ventricular fibrillation

The only effective therapy is the use of a defibrillator for electric cardioversion applying 200–400 J.

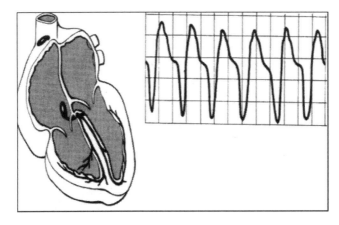

Fig. 44 Cross-section of the heart with the atrial pacemaker and the AV-node, both of which are involved in cocaine-related tachyarrhythmia, a premonitory sign of ventricular fibrillation

Treatment of Hypertension

- Na-Nitroprusside infusion, or
- Urapidil (25–50 mg), or
- Nitrogylcerine infusion or
- Glycerolnitrate Spray or
- Nifedepin sl (5–10 mg) or
- Clonidine (0.15 mg slowly iv)

Treatment of Hyperthermia

- Cold infusions
- Ice cubes axilla/groin
- Dantrolene 2.5 mg/kg repetitively

Treatment of Seizures

- Diazepam (10–20 mg), or
- Midazolam (5–15 mg), or
- Phenobarbital (50–100 mg), or
- Clonazepam (1–2 mg)

> **Since there is no specific antidote available therapy consists of treatment of symptoms until the acute effects are gone**

Lethal Doses of Cocaine

There's a lot of debate, and not a lot of data, about the lethal dose of cocaine. The lack of uniformity seems to have much to do with the extreme variation in individual response to cocaine. The classic *fatal* dose has been reported to be 1.2 g (orally), but there are no studies to underline this. Anecdotal reports from ER's and coroners suggest, that death from a single intranasal dose of 20 mg may ensue. Viewed from the perspective that the minimum dose required to get either a subjective *high* or objective findings of increased blood pressure, etc. is 25 mg, makes this all the more alarming. Individuals with defects in their ability to neutralize cocaine (pseudocholinesterase deficiency) may suffer death from relatively small doses. Since smokers are rarely able to regulate their intake (some freebase binges last for days), these abusers are at very high risk for overt toxicity.

Report from emergency departments (EDs) are typical of what is been seen by paramedics and other emergency department personnel many times often late at night. The report of a "witnessed seizure" is reminiscent in which a young athlete was reported to become agitated, with his eyes rolling back and observed to be vomiting and foaming at the mouth. In this case, the paramedics were unable to help him while in other cases, the outcome was more favorable. Typically these patients are commonly seen after exhibiting a seizure, or a hypertensive crisis. It must always be kept in mind, that aside from acute effects, there is also a long-term damage being done to the transmitter system.

Chronic Toxicity of Cocaine use

Once the patient survived the symptoms of an overdose, what can they look forward to next? Cocaine, even for those who feel that they don't use that much or that often, has some chronic effects that are not to be scoffed at. Paranoia is a consequence of alterations in dopamine activity, particularly in the limbic system. It is well known that other CNS stimulants (amphetamines, methylphenidate, etc.) can cause psychotic manifestations. Cocaine taken over some time, and in varying amounts, can lead to behavioral abnormalities, which manifest themselves most frequently as suspiciousness, hostility, delusions, hallucinations, etc. For example, many "regular" users of cocaine carry firearms, which can lead to visits to the emergency room (ER) secondary to shot wounds acquired in violent attacks.

An especially bizarre effect of cocaine are the so-called "Coke Bugs", altered tactile sensation (tactile hallucination) that bugs are crawling under one's skin (Fig. 45). This can lead to a person actually picking imaginary bugs off their arms and legs with tweezers and such (self excoriation). It is recorded that individuals suffering from this hallucination have produced *proof* of the existence of the *"bugs"* by bringing little vials of them into the ER of course when examined the "bugs" turn out to be pieces of dead skin.

Much more commonly, chronic users of cocaine are unaware of the long-term damage being done to their bodily part. A prime example is the recent finding in animal studies of cardiac enzyme depletion. While not adequately evaluated in humans yet, the consequences may be that younger users are setting themselves up for serious cardiac compromise in the future. Some pathologists at Stanford looked at autopsy findings of persons who had died of cocaine-associated sudden death. Their findings were startling as they discovered that cocaine users demonstrated *wide-spread, intense, contraction band necrosis*, and they felt that death resulted when this necrosis supplied the *anatomic substrate for a malignant type re-entry arrhythmia* [50, 51].

Once the recreational user turns into a chronic user, brain lesions in PET studies are demonstrated. This is accompanied by abnormalities in the dopaminergic system with an increased release in the metabolite homovanillic acid (HVA), which is accompanied by symptoms of Parkinson and tremor suggesting a deficit in the transmitter dopamine [52].

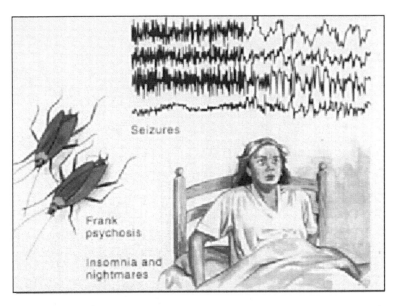

Fig. 45 Effects of chronic abuse of cocaine with seizures, tactile paranoia, visual hallucinations and insomnia

The entry of cocaine molecules into the system causes an abnormal increase of neurotransmitters at the synapse, leading to an overstimulation of the reward areas and a feeling of intense euphoria. This state may be accompanied by paranoia and hyperexcitability. With continued use, cocaine molecules further interfere with the release of neurotransmitters and block the receptor sites in the reward area. At the same time, the number of receptor sites increases, resulting in an even greater discrepancy between the amount of transmitter available and the number of receptor sites occupied (Fig. 46). Since cocaine also works by stimulating the release of neurotransmitters, and they are now short in supply, more cocaine is required to obtain a *high*. Craving remains at a high level, and the user experiences a generalized feeling of depression and unhappiness. Eventually, these symptoms become severe and terminate in the *crash*, which is characterized by intense craving, insomnia, restlessness, and anhedonia (profound depression, total absence of pleasure, and increased feelings of worthlessness).

Lastly, problems with the nasal passages are common, sometimes 'requiring surgery for repair of perforated septa, and finally, for the IV users, there is the potential of acquiring AIDS.

A rather interesting and not fully understood effect of repeated cocaine use is called *kindling*.' This refers to a phenomenon of *reverse tolerance* that has been seen in the laboratory, but not yet confirmed in humans [53]. It seems that repetitive, subthreshold electrical stimulation of the limbic system, which relied on dopamine as a neurotransmitter, can lead to sensitivity to catecholamines, and lower the

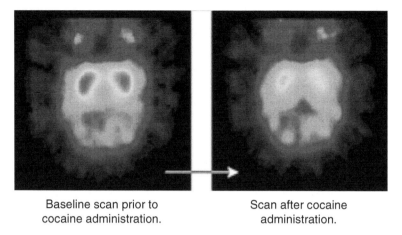

| Baseline scan prior to cocaine administration. | Scan after cocaine administration. |

Fig. 46 Positron Emission Tomography (PET) scans showing the average level of dopamine receptors in primates' brains. Red is high- and blue is low-concentration of dopamine receptors. The higher the level of dopamine, the fewer receptors there are. This chronic use results in an increase of receptors, however, with less dopamine being available for binding

seizure threshold. This may help to explain why the next dose, even if it is less than that usually taken, might be the one that kills the user. In this respect it is hypothesized that carbamazepine (CBZ), which is an anticonvulsant medication has a potential as a treatment for cocaine abuse because of its ability to block cocaine-induced "kindling" in rodents [54]. Since kindling has also been postulated to be a model for the neurophysiological basis of cocaine craving, CBZ may reverse the DA receptor supersensitivity that purportedly is the result from chronic cocaine use, and its potential as a treatment for cocaine dependence.

Summary of Acute-Chronic Effects of Cocaine

During Early Use

- Magnification of pleasure, euphoria
- Alertness and in some cases hyperalertness
- Increased and/or a (grandiose) sense of well being
- Decreased anxiety
- Lower social inhibitions: more sociable and talkative
- Heightened energy, self-esteem, sexuality and emotions aroused by interpersonal experiences
- Appetite loss; weight loss

With Compulsive Use

- Extreme euphoria – "mental orgasm"
- Uninhibited behavior pattern
- Impaired judgment
- Feeling of grandiosity
- Impulsive reaction
- Hypersexuality
- Hypervigilance
- Compulsive actions
- Extreme psychomotor activation/agitation
- Anxiety; irritability; argumentative
- Transient panic
- Paranoia
- Visual hallucinations
- Gustatory and auditory pseudohallucinations

- Altered tactile sensations (with excoriation)
- Tactile paranoia ("coke bugs")
- Terror of impending death
- Poor reality testing; delusions
- Extreme weight loss

Physical Effects of Chronic Abuse

- Chronic sore throat
- Subthreshold seizure levels ("kindling")
- Nasal septal defect
- Contraction band necrosis in the heart
- Cardiac enzyme depletion
- Hoarseness
- Shortness of breath
- Bronchitis
- Lung cancer (smoking "ctack")
- Pulmonary edema
- Emphysema and other lung damage
- Respiratory problems such as congestion of the lungs, wheezing, and spitting up black phlegm especially after crack
- Burning of the lips, tongue, and throat
- Slowed digestion
- Weight loss
- High incidence of dependence
- Blood vessel constriction
- Increased blood pressure
- Increased heart rate
- Brain seizures that can result in suffocation
- Dilated pupils
- Sweating
- Rise in blood sugar levels and body temperature
- Suppressed desire for food, sex, friends, family, social contacts
- Heart attack
- Stroke
- Death

> **Emotional/Psychological Effects**
> - Sadness and depression
> - Loss of interest in appearance
> - Unexplained loss of household valuables or vanishing of cash due to the expense of the drug
> - Sleeplessness
> - Extreme paranoia
> - Intense craving of the drug
> - Schizophrenic-like psychosis with delusions and hallucinations

Duration of Withdrawal Symptoms Following Cocaine Abuse

2–48 h	2–10 weeks
Paranoia	Depression
Suicidal attempts due to	Depression, Anxiety
Excessive sleep,	Craving
→ Relapse!	→ Relapse!

> **Craving for the drug, even following months of abstinence, is a major reason why a previously stable former cocaine user has a relapse [55]**

Cocaine Use in Pregnancy

Women addicted to cocaine often continue drug use through pregnancy, despite risks to the fetuses they are carrying. Primate studies have shown that intrauterine cocaine exposure (during a period corresponding to the second trimester in humans) results in a decrease in the number of neurons in the cerebral cortex and disorganization of the normal laminar structure of the cortex [56, 57]. Even brief exposure at a particularly vulnerable time in brain development may have lasting deleterious effects of greater magnitude than greater exposures at other times [58]. Moreover, the postnatal age at which the effects of cocaine are measured, whether in humans or animals, may show evolving outcomes [59]. In humans, the attribution of outcomes to drug effect is complicated by the observation that the circumstances under which children have been raised subsequent to cocaine exposure affect their behavior [60].

Cocaine and Pregnancy

* Increased incidence of stillbirths
* Increased incidence of miscarriages
* Premature (often fatal) labor and delivery
* In males, cocaine attaches to the sperm causing damage to the cells of the fetus.

Effects of Cocaine on the Fetus

* Seizures or strokes
* Cerebral palsy
* Mental retardation
* Vision and hearing impairments
* Urinary tract abnormalities
* Autism and learning disabilities
* Babies exposed to cocaine experience painful and life threatening withdrawal, are irritable, have poor ability to regulate their own body temperature and blood sugar and are at increased risk of having seizures.

Treatment Options in Cocaine Abuse

Since the cocaine user is prone for a relapse the primary goal in therapy is the prevention of further use. Within this frame, treatments for drug abuse are continually improving. For instance, for twenty-five years methadone was the only treatment for opiate dependence. Although LAAM (α-levoacetylmethadol) was removed from the European market due to reports of severe cardiac-related adverse events, and in 2003 Roxane Laboratories Inc. discontinued the product ORLAAM™ in the US. Presently, buprenorphine (Subutex®) and naltrexone (Trexan®, Revia®) are being used as treatment options [60]. Similarly, disulfiram, naltrexone and/or amprosate are being prescribed for alcohol dependence [61]. However, the treatment of cocaine abuse poses a more difficult challenge for addiction pharmacotherapy. In future, potential treatments may be based on immunopharmacotherapeutic agents that can suppress the effect of cocaine on behavioral and locomotive actions. At the horizon a second-generation vaccine is emerging, which protects against the psychoactive effects of cocaine [62]. Similarly a catalytic monoclonal antibody (mAB) is under development to bind and degrade cocaine by hydrolysing the benzoate ester of cocaine [63].

The actual mechanism whereby cocaine produces euphoria is still unknown. Presently, Sigma 1 receptors are being discussed as they are unique endoplasmic reticulum proteins that bind certain steroids, neuroleptics and psychotropic drugs. Sigma 1 receptors form a trimeric complex with ankyrin B and IP2R type 3 in NG-108 cells, which regulate Ca^{2+}-signalling may represent an active site for the binding of cocaine and neurosteroids [64].

Basically, however, there is a sequence of different steps in the treatment, which can be outlined as follows

1. Detoxification
2. Structured cocaine-free environment
3. Hospitalization or out-patient therapy
4. Pharmacologic therapy

Note, that this is treatment of abuse, not necessarily of "addiction". It is hoped that abusive behavior can be curtailed before it develops into a full blown behavioral pattern. The elements of a program will vary according to the philosophy of the treatment organization, but most of those listed here will likely be included.

Detoxification

All too often this word means *cure* for many illicit users. As discussed in the previous chapters, the mere removal of the substance from the user is not sufficient to extinguish craving. While detoxification is essential, it's only a first step in a rather lengthy process of recovery.

Cocaine-Free Environment

There are many factors that predispose individuals to use, and environment is certainly high on the list. To expose the recovering abuser to the same conditions that got them into use in the first place is to lose from the very start. If others in the patient's working or living arrangements are users, it will be difficult to extinguish craving, as there will be easy access to the substance when "drug hunger" calls.

Hospitalization or Outpatient Therapy

Depending upon the various factors (economics, environmental, psychological, physical) that apply, a decision must be made about the need for more or less supervision during detoxification and recovery.

Pharmacological Therapy

With new research adding to our understanding of cocaine's mechanism of action, pharmacologic intervention may prove to be of immense benefit in restoring the chemical imbalances caused by cocaine abuse. Pharmacologic therapy in cocaine addiction, however should not be confused with other pharmacologic approaches to dealing with drug abuse, such as methadone in opioid "replacement" therapy, or Trexan® (i.e. naltrexone, an opioid antagonist) in the rehabilitation phase. The latter compounds are used in the treatment of opioid dependency, and are directed at maintaining the physical state of the opioid addict, or preventing re-addiction of those addicts. Drugs used in the treatment of cocaine dependency are directed towards normalizing the chemical imbalances caused by cocaine, and reducing the craving for that drug. According to an early neurological model of addiction, cocaine-seeking behavior results from a deficit or imbalance of neurochemicals (neurotransmitters and neuromodulators), particularly dopamine. With the primary underlying deficit being depression, probably caused by depletion of dopamine, the target of drug therapy is either to treat the depression directly, or to restore the activity of the depleted neurotransmitter dopamine, since this neurotransmitter serves as a target messenger in the limbic system of the brain, causing feelings of reward. Under the traditional, though limited, assumption that dopamine depletion is the key to cocaine addiction, pharmacotherapies were developed using three alternate modes of dopamine substitution or restoration:

1. Substances that mimic the natural action of dopamine at its receptor sites in the reward area, such as bromocriptine mesylate or similar agents.
2. Substances that release dopamine, such as amantadine hydrochloride.
3. Substances that are used to synthesize dopamine, e.g., precursors, such as levodopa and L-tyrosine.

Aside from pharmacological treatment, restoration of neurotransmitter deficits with nutritional supplements seems a rational approach. By increasing the synthesis of neurotransmitters that are in short supply, they facilitate the release of the stimulant neurotransmitter dopamine, prevent the breakdown of enkephalin, and allow natural processes to stimulate the reward sites, leading to feelings of well-being. One promising approach to restoration of the neurotransmitter deficit involves precursor amino acids, and their production requires certain vitamins and minerals. While amino acid supplements have a long and useful history in the treatment of substance abuse, the "reward cascade" suggests that a specific mixture of amino acids and vitamins would be even more beneficial. One question in the use of amino acid precursors is the ability of these compounds to reach the brain from the blood in sufficient amounts. Large neutral amino acids (LNAAs), such as tryptophan, tyrosine, and phenylalanine, are transported across the blood-brain-barrier (BBB) by a special carrier. The BBB normally serves to isolate, to a large degree, fluids of the brain from compounds dissolved in blood plasma. There are a number of circumstances that can alter the action or effect of the LNAA carrier or of the BBB. Diet, stress, and drugs are three such factors. LNAA transport to the brain is virtually in direct relation to the concentration in the blood. Carbohydrate and protein rich diets have very different effects on blood LNAA concentrations. A carbohydrate meal increases insulin secretion. Insulin moves certain LNAAs to muscle and changes the amount of other LNAAs bound to blood proteins. Thus, carbohydrates increase the amount of tryptophan, tyrosine, and phenylalanine that reach the brain. A protein rich meal has the opposite effect.

L-Tryptophan

This amino acid is a food supplement and freely available in health food stores. Being an essential amino acid it means that it cannot be synthesized by the organism and therefore must be part of its diet. It is a precursor of serotonin and melatonin (Fig. 47), and may aid in the replenishment of that neurotransmitter. Clinical research confirmed tryptophan's effectiveness as a sleep aid [65, 66] and for a growing variety of other conditions typically associated with low serotonin levels or activity in the brain such as seasonal affective disorder [67]. In particular, tryptophan has been showing considerable promise as an antidepressant alone and as an "augmenter" of antidepressant drugs [68], although the reliability of these clinical trials has been questioned [69].

Also, the metabolite of tryptophan, i.e. 5-hydroxytryptophan (5-HTP), readily crosses the blood-brain barrier where it is rapidly decarboxylated to serotonin (5-hydroxytryptamine or 5-HT) [70] thus being suggested as a treatment for depression [70]. However, serotonin has a relatively short half-life since it is rapidly metabolized by monoamine oxidase, and therefore is likely to have limited efficacy.

L-Tyrosine

Similar to the previous amino acid this also serves as a precursor of dopamine synthesis, which may aid in the replenishment of that neurotransmitter (Fig. 48).

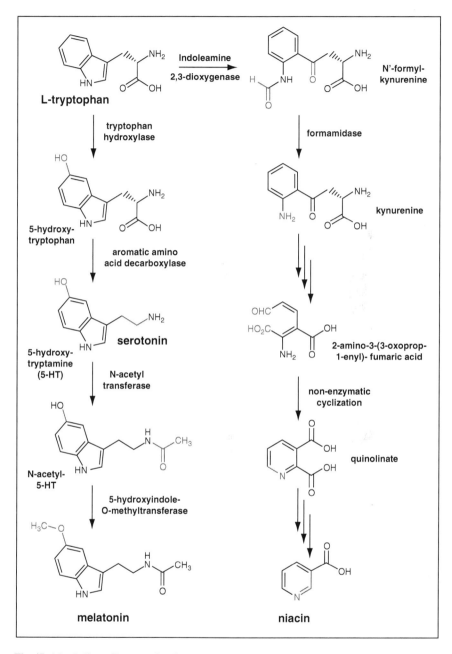

Fig. 47 Metabolism of L-tryptophan into serotonin, melatonin (*left*) and niacin (*right*). Transformed functional groups after each chemical reaction are highlighted in red.

In the adrenal gland, tyrosine is converted to levodopa by the enzyme tyrosine hydroxylase (TH). TH is also the rate-limiting enzyme involved in the synthesis of the catecholamine hormones dopamine, norepinephrine (noradrenaline), and

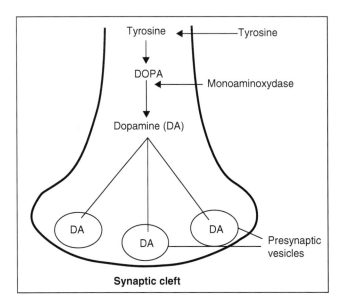

Fig. 48 Tyrosine being an essential amino acid in the synthesis of dopamine to relieve depression

epinephrine. Although tyrosine does not seem to have any significant effect on mood, cognitive or physical performance under normal circumstances, however, an effect on mood is more noticeable in humans subjected to stressful conditions [71–74]. The recommended daily dosage of 10 mg/kg for a body weight usually amounts to 500–1500 mg per day in the adult [71].

Use of Antidepressants to Relieve Post-cocaine Depression

The drugs of choice in the treatment of depression, the tricyclics, alter serotonin activity, an important neurotransmitter affected by cocaine, and relieve distressful symptoms. However, they are potentially cardiotoxic on their own, and have some other bothersome side effects such as heart rhythm disturbances, hypotonia, impaired liver function, weight gain, and/or bone marrow depression. These series of antidepressants are primarily designed to restore the balance of dopamine to functional levels. Under the assumption that depression is a contributory problem in cocaine addiction, monoamine oxidase inhibitors (e.g. phenelzine) or tricyclic anti-depressants (e.g. imipramine or desipramine) are used. These drugs not only interfere with the breakdown of dopamine, but reduce the breakdown of norepinephrine, a transmitter involved in depression.

Bupropion

Previously known as amfebutamone (Wellbutrin®, Zyban®, Budeprion® and Buproban®) is an atypical antidepressant that acts as a norepinephrine and dopamine re-uptake inhibitor and a nicotinic antagonist. Bupropion in general belongs to the chemical class of aminoketones and is similar in structure to the stimulant cathinone, to the anorectic diethylpropion, and to phenethylamines. Initially researched and marketed as an antidepressant, bupropion was subsequently found to be effective as a smoking cessation aid. In 2006 it was the fourth-most prescribed antidepressant in the United States retail market, with more than 21 million prescriptions. Bupropion lowers seizure threshold and its potential to cause seizures was widely publicized. However, at the recommended dose the risk of seizures is comparable to the one observed for other antidepressants. In contrast to many psychiatric drugs, bupropion is an antidepressant of the newer generation with lesser side effects. It does not cause weight gain or sexual dysfunction, and presently has been approved by the US FDA for the treatment of depression ("cocaine blues") associated with ending cocaine use.

Phenelzine

Under the proprietary name Nardil®, this agent appears to be effective with cocaine patients in reducing sleep disturbances and depression, enhancing concentration, and augmenting energy, but requires several weeks before results are obtained. If the patient relapses during this time, the interaction between cocaine and phenelzine can cause a fatal hypertensive crisis due to adrenalin surge.

Desipramine (Norpramin®, Pertofrane®)

This tricyclic antidepressant agent was one of the first medications to be studied as a treatment for cocaine dependence and, as such, is one of the most extensively studied pharmacotherapies for cocaine dependence to date [75]. The rational for its use is that it blocks re-uptake of norepinephrine and to a lesser extent dopamine, thus acting as a specific antianhedonic agent in cocaine-dependent patients, and similar to other antidepressants, it seems to hold the greatest promise in treating depression, even though having a slow mode of action. It seems to ameliorate dysphoria and anhedonia during the post-cocaine crash. Controlled clinical results, however, raise some questions about the efficacy of tricyclic antidepressants in reducing both cocaine craving and usage.

Dopamine Antagonists

Studies have indicated that haloperidol (Haldol®) and similar dopamine (DA) antagonists (e.g. Flupentiol®) might be contraindicated for longterm treatment of cocaine abuse. It appears that one possibility is that such agents could contribute to enhanced cocaine effects [76]. These findings may also help to partially explain the high prevalence of cocaine abuse in neuroleptic-maintained schizophrenics [77].

Bromocriptine (Parlodel®)

While it is postulated that chronic cocaine use may deplete central dopamine (DA) stores, this could result in supersensitivity of dopaminergic receptors. Thus DA hypofunction induced by cocaine abuse may underlie craving and withdrawal symptoms often observed in recently abstinent cocaine dependent patients. Treatment with bromocriptine, a second generation antidepressant, therefore might reverse dopaminergic deficits induced by cocaine and ameliorate craving and withdrawal. Bromocriptine is an agonist with high affinity for the D2 receptor by directly stimulating the postsynaptic dopamine receptors, thus by-passing the depleted presynaptic dopamine vesicles. It reduces craving for cocaine in certain patients within minutes after administration. However, this drug has well-known side effects, including headaches, sedation, tremor, vertigo, and dry mouth. Furthermore, bromocriptine may itself create drug dependency since it has been demonstrated to be self-administered by laboratory animals. Other side-effects contain a high incidence of gastrointestinal upset with nausea and vomiting and psychotic symptoms. Given its DA agonist properties, this drug is considered for further clinical trials to assess its efficacy for treatment of primary cocaine dependence.

Amantadine hydrochloride (Symmetrel®)

One of the newest compounds to be evaluated in the treatment of cocaine withdrawal reactions is amantadine. This agent increases dopaminergic transmission, but whether the mechanism is dopamine (DA) release, directly effects DA receptors, or DA re-uptake blockade is unclear. Patients receiving this drug do experience, from time to time, Parkinsonian symptoms such as muscular rigidity or impaired reflexes. The most important problems are that the patient exhibits severe dopamine depletion, and the dopamine releaser does not work because there is no dopamine to release. It has been shown to be effective in reducing depression and craving for cocaine when used in combination with the amino acids L-tryptophan and L-tyrosine. A comparison of amantadine and bromocriptine has been published [78], which shows significant improvements and less adverse reactions than seen with bromocriptine.

L-Deprenyl (Seligiline®)

It is an irreversible monoamine oxidase type B inhibitor that specifically inhibits the metabolism of DA. Its present indication is for the treatment of Parkinson's disease. The ability of L-deprenyl to potentiate DA has led to consideration of its use in the treatment of cocaine dependence. In studies, however, there was no difference in cardiovascular effects or drug *liking* when using the cocaine-alone administration or the L-deprenyl-cocaine combination.

Methylphenidate (Ritalin®, Rilatine®, Attenta®, Methylin®, Penid®, Rubifen®)

This drug is a stimulant drug primarily used in the treatment of childhood attention deficit hyperactivity disorder (ADHD). Methylphenidate is a dopamine (DA) agonist with pharmacological properties that include DA release and re-uptake inhibition. A study demonstrated a novel approach to drug development and showed that this class of medications may be useful in the treatment of cocaine dependence [79]. Focalin® is a newer preparation containing only the dextro-methylphenidate, rather than the usual racemic dextro-and levo-methylphenidate mixture of other formulations. A recent way of taking methylphenidate is by using a transdermal patch (Daytrana®).

L-Dopa Ligands (Dopar®, Larodopar®, and Sinemet®)

This is a combination of carbidopa and levodopa, and all dopa ligands are used as a prodrug to increase dopamine levels for the treatment of Parkinson's disease, since it is able to cross the blood–brain barrier, whereas dopamine itself cannot. Once levodopa has entered the central nervous system, it is metabolized to dopamine by aromatic L-amino acid decarboxylase (Fig. 49). Pyridoxal phosphate (vitamine B6) is a required cofactor for this decarboxylation, and may be administered

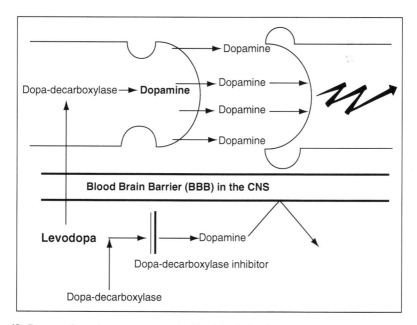

Fig. 49 Because dopamine cannot cross the blood–brain–barrier (BBB) the precursor levodopa is administered, which in combination with an inhibitor of the enzyme decarboxylase, is able to diffuse into the CNS where it is converted into dopamine thus replenishing the deficient vesicles at the nerve terminals

along with levodopa, usually as pyridoxine. However, conversion to dopamine also occurs in the peripheral tissues, i.e. outside the brain. This is the primary mechanism of the adverse effects of levodopa. It is standard clinical practice to co-administer a peripheral DOPA decarboxylase inhibitor – carbidopa or benserazide – and often a catechol-O-methyl transferase (COMT) inhibitor, to prevent synthesis of dopamine in peripheral tissue. Co-administration of pyridoxine without a decarboxylase inhibitor accelerates the extracerebral decarboxylation to such an extent that it cancels the effects of levodopa administration, a circumstance, which historically caused great confusion. For years, these agents were used to treat Parkinson patients by replenishing their supply of dopamine presynaptically. They are less satisfactory in reducing cocaine craving than bromocriptine or amantadine. Furthermore, levodopa agents can produce hallucinations, and they have many of the same side effects as seen with bromocriptine.

Lithium (Lithobid®, Lithonate®, Lithotabs®, Eskalith®)

This compound generally is used in the treatment of bipolar manic-depressive illness. It has been studied as a potential blocker (antagonist) of cocaine effects, especially euphoria, thus avoiding relapse.

Cocaine Antagonists

A variety of medications have been examined for their effectiveness in blocking the reinforcing effects of cocaine. These drugs, including mazindol, fluoxetine, carbamazepine, naltrexone, disulfiram, and the highly potent antagonist 4-iodococaine, have been the subject of study over the past several years. They all have a broad range of pharmacological properties, and differ greatly in primary indication. However, all have been postulated to antagonize the effects of cocaine through pharmacological properties specific to each drug, which might alter neurobiological and reinforcing effects of cocaine. Comparative studies are still pending.

Non-pharmacological Treatment of Cocaine Addiction

Condition Behavioral Therapy (CBT)

Aside from phramacological treatment and based on social learning theory, condition behavioral therapy or CBT, presents another option. Within this theory it is assumed that conditioning presents an important factor in how individuals begin to use and abuse substances and that they learn to do so. The several ways individuals may learn to use drugs include modeling, operant conditioning, and classical conditioning.

Such conditioning can be derived from Pavlov's classical experiments. Pavlov demonstrated that, over time, repeated pairings of one stimulus (e.g., a bell ringing) with another (e.g. the presentation of food) could elicit a response (e.g. a dog salivating). Over time, cocaine abuse may become paired with money or cocaine paraphernalia, particular places (bars, places to buy drugs), particular people (drug-using associates, dealers), times of day or week (after work, weekends), feeling states (lonely, bored), etc. Eventually, exposure to those cues alone is sufficient to elicit very intense cravings or urges that are often followed by cocaine abuse. For each instance of cocaine use during treatment, the therapist and patient do a functional analysis, that is, they identify the patient's thoughts, feelings, and circumstances before and after the cocaine use. Early in treatment, the functional analysis plays a critical role in helping the patient and therapist assess the determinants, or high-risk situations, that are likely to lead to cocaine use and provides insights into some of the reasons the individual may be using cocaine (e.g., to cope with interpersonal difficulties, to experience risk or euphoria not otherwise available in the patient's life). Later in treatment, functional analyses of episodes of cocaine use may identify those situations or states in which the individual still has difficulty coping. CBT can be thought of as a highly individualized training program that helps cocaine abusers unlearn old habits associated with cocaine abuse and learn or relearn healthier skills and habits. By the time the level of substance use is severe enough to warrant treatment, patients are likely to be using cocaine as their single means of coping with a wide range of interpersonal and intrapersonal problems.

New Options in the Treatment of Cocaine Dependency

The Antiepileptic Gamma Vinyl-GABA (GVG)

The National Institutes of Health (NIH) suggests that gamma vinyl-GABA (GVG), a drug used to treat epilepsy, may prove to be an effective treatment for cocaine addiction. Researchers from New York University School of Medicine and Brookhaven National Laboratory in Upton, New York, reported in a small, preliminary clinical trial conducted in Mexico that this drug could cut cocaine use dramatically in people who had used cocaine daily for at least 3 years [80]. GVG reduced levels of dopamine, the "*feel-good*" chemical that floods the brains of cocaine users, providing the "*high*" they crave (Fig. 50). Using GVG to temper the dopamine system may very effectively block the addiction-related effects of cocaine. The people who completed the study said their craving for the drug was eliminated after 2–3 weeks of continuous GVG administration, the authors report. In addition, those who completed the trial also showed improved self-esteem, re-established healthy family relationships, went to work, or actively sought work. Such preliminary finding has important implications for our medications development program.

GVG, also known as Vigabatrin®, is approved in many countries as a treatment for epilepsy. Its main effect is that it increases the amount of another brain chemical involved in nerve cell communication, GABA, and thus helps moderate seizures. The US Food and Drug Administration has not approved GVG for treating epilepsy or any indication because of concerns about the relatively high incidence of tunnel vision that has occurred in people given the drug over many months or years. Vigabatrin® (or CPP-109, Catalyst Pharmaceutical Partners/USA) has been tested in regard to potential side effects, especially on vision impairment, and none of the eight people who completed the pilot study reported vision changes of any kind.

Cocaine normally results in an increased concentration of dopamine in the brain causing the "high" or exaggerated sense of pleasure associated with drug abuse. This is where CPP-109 sets in by indirectly lowering the level of dopamine in the brain via GABA (gamma-aminobutyric acid), is a neurotransmitter in the brain, which inhibits the release of dopamine. GABA, however, is broken down by GABA transaminase (GABA-T). CPP-109 works by inhibiting GABA-T and

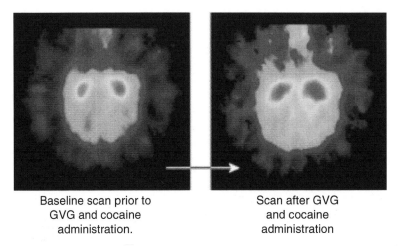

| Baseline scan prior to GVG and cocaine administration. | Scan after GVG and cocaine administration |

Fig. 50 GVG increases the amount of the neurotransmitter GABA in the brain and reduces the level of dopamine in the region of the brain that is thought to be involved in addiction

consequently increasing the level of GABA, which then lowers the level of dopamine and turns off the "*high*". CPP-109 seems to work without the apparent side effects typically associated with GABA agonists. Thus targeting brain GABAergic systems with drugs such as CPP-109 is a potentially effective treatment for cocaine, methamphetamine and other substance dependencies. Previous studies in established animal models of addiction, involving both rats and primates, have shown that Vigabatrin® interrupts the neural mechanisms essential for addiction (Fig. 50). In preclinical studies, Vigabatrin® prevented the characteristic drug-seeking behavior of addicted animals. Meanwhile three human trials of Vigabatrin® have been completed in Mexico with patients addicted to cocaine or methamphetamine, including a US Phase II, double-blind, placebo-controlled trial in 103 patients, which was completed in 2007. Data from these three trials provide clinical evidence of Vigabatrin®'s potential as a safe and effective treatment for patients with these addictions (Fig. 51). By 2008, the company Catalyst initiated enrollment of patients for a US Phase II 180-patient, multicenter, double-blind, placebo-controlled clinical trial evaluating the use of CPP-109 in treating patients with cocaine addiction.

Disulfiram (Antabuse®)

Disulfiram is a drug normally used to help patients with alcohol disorders to remain sober. Patients taking disulfiram and who ingest even small amounts of alcohol develop a reaction that produces nausea, flushing, vomiting, and throbbing headache. At times, this reaction can be severe and can lead to critical illness, such as severe respiratory or circulatory problems. In combination with cognitive behavioral

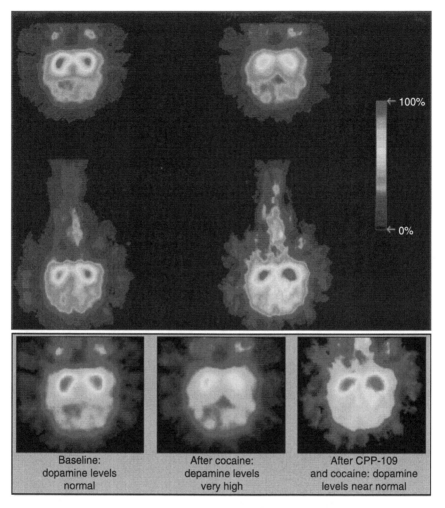

Fig. 51 Composite brain scans on receptor availability before, during cocaine use, and following treatment with CPP-109 in the corpus striatum of baboons (red indicates higher availability vs. magenta indicating a lower receptor availability). Cocaine decreases receptor availability (*top right*) compared to baseline values (*top left*). When pretreated with GVG, however, cocaine did not alter receptor availability (*lower right*) compared to baseline values (*lower left*)

therapy (CBT) disulfiram appears to be effective in reducing cocaine use, especially among cocaine users who are not dependent on alcohol. Since alcohol is a powerful "cue" for using cocaine, and can impair judgment and lower resistance to cravings for cocaine, researchers hypothesize that by reducing alcohol use with disulfiram, users might be less likely to abuse cocaine. However, use of disulfiram has not been evaluated in general populations of cocaine users. A study in randomly 121 cocaine-dependent adults (average age 34.6 years) were assigned to receive either disulfiram or placebo over a 12-week period. Participants were also randomized to participate in either cognitive behavioral therapy (CBT) or interpersonal psychotherapy (IPT), a less structured form of behavioral therapy. Participants assigned to disulfiram reduced their cocaine use significantly more than those assigned to placebo, and those assigned to CBT reduced their cocaine use significantly more than those assigned to IPT [81]. This first placebo-controlled trial demonstrates that disulfiram therapy is especially effective for nonalcoholic cocaine users, as the effects of disulfiram treatment were most pronounced in participants who did not meet the criteria for current alcohol abuse or dependence and in those who abstained from alcohol during the trial.

Use of an Anti-cocaine Vaccine

Normally, the dopamine D2-receptor receives signals in the brain triggered by dopamine, a neurotransmitter needed to experience feelings of pleasure and reward. Without receptors for dopamine, these signals get "jammed" and the pleasure response is blunted. Previous studies in animals have shown that chronic abuse of alcohol and other addictive drugs increases the brain's production of dopamine. Over time, however, these drugs deplete the brain's D2-receptors and rewire the brain so that normal pleasurable activities that stimulate these pathways no longer work, leaving the addictive drug as the only way to achieve this stimulation. A study demonstrating dramatic reductions in alcohol use in alcohol-preferring rats infused with a gene inducing dopamine D2-receptor activation hypothesized that the same would hold true for cocaine-dependent individuals, who may have their need for cocaine decreased if their D2 levels are boosted. Researchers tested this hypothesis by injecting a virus that had been rendered harmless and altered to carry the D2-receptor gene directly into the brains of experimental rats that were trained to self-administer cocaine, a technique used in the earlier alcohol study. The virus acted as a mechanism to deliver the gene to the nucleus accumbens, the brain's pleasure center, enabling the cells in this brain region to make receptor proteins themselves. An animal study examined how the injected genes affected the rats' cocaine-using behavior after they had been taking cocaine for 2 weeks. Having received the D2 receptor treatment, the rats showed a 75% decrease in self-administration of the drug. This effect lasted 6 days before their cocaine self-administration returned to previous levels.

Vaccination of the Cocaine User

TA-CD is an active vaccine developed by the Xenova Group/USA, which is used to negate the effects of cocaine, making it suitable for use in the treatment of addiction, and is made by combining norcocaine with inactivated cholera toxin. It works in much the same way as a regular vaccine. A large protein molecule attaches to cocaine, stimulating the response for antibodies, which then destroy the molecule.

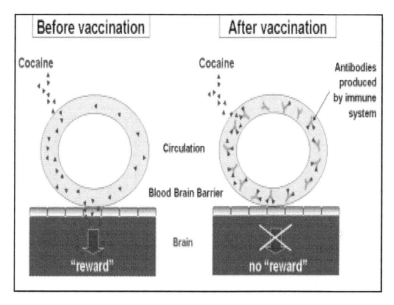

Fig. 52 Principles of vaccination preventing further use of cocaine and avoiding relapse

This also prevents the cocaine from crossing the blood-brain barrier, negating the euphoric high and rewarding effect of cocaine caused from stimulation of dopamine release in the mesolimbic reward pathway (Fig. 52). This anti-cocaine vaccine is in early testing stages. Reports indicate that the vaccine successfully mounted an antibody response against free cocaine in the blood that lasted nearly 3 months. It is believed that antibodies to other stimulants might also be developed [62]. In addition, TA-CD also has been shown to lower the effect of cocaine on the heart.

Newer Dopamine Reuptake Inhibitors

GBR-12909 (Vanoxerine)

Vanoxerine is a selective dopamine uptake inhibitor. Because of this action, it reduces cocaine's effect on the brain, and may help to treat long-term cocaine addiction. Preliminary studies have shown that GBR 12909, when given to primates, suppresses cocaine self-administration. As of 2006 clinical studies phase 2 are in progress.

Venlafaxine (Effexor®) and Ecopipam

Although not a dopamine reuptake inhibitor, venlafaxine is a serotonin/norepinephrine reuptake inhibitor that has been successfully used to combat depression caused by cocaine withdrawal and to a lesser extent, the addiction associated with the drug itself. Further clinical studies, especially in regard to side-effects are planned to prove its usefulness.

Recently reported research demonstrated that the selective dopamine D1/D5 receptor antagonist ecopipam diminished the euphoric and anxiogenic effects of cocaine, and stemmed craving [82]. It has been suggested that low doses of the antagonists might be useful in blocking relapse or blunting effects of relapse when it occurs. D3 receptor partial agonists also have been effective in reducing cocaine craving, so a combination of drugs acting on D1 and D3 receptors might prove particularly powerful for treating cocaine addiction [83].

Outlook Regarding Therapy in Cocaine Abuse

One of the biggest challenges in treating drug addiction is the danger of relapse after quitting. Drug-associated cues can trigger persistent drug cravings that grow stronger the longer the abstinence, but what mediates the increasing reactivity of the brain to these cues is not well understood. It is known that cocaine-seeking also depends on activation of a subgroup of glutamatergic receptors in the nucleus accumbens. Recently researchers have shown that rats that undergo prolonged withdrawal from cocaine show an increase in the numbers of this type of synaptic APMA (α-amino-3-hydroxy-5-methyl-4-isoxazol-propionic acid) receptors. This, in return, leads to the increased reactivity of nucleus accumbens neurons to cocaine-related cues and an increase in drug-seeking by the animals. Thus, the AMPA receptors in the nucleus accumbens present a new approach to tackle the problem of cocaine-related relapse [84].

References

1. Stocker S. Cocaine abuse may lead to strokes and mental deficits. NIDA Notes. 1998; 13:10–2.
2. Sturgess R. Freud, Sherlock Holmes and Coca-Cola – the cocaine connection. Pharmacol J. 2000;265:915–7.
3. Wilson LD, Jeromin G, Shelat C, Huettl B. Tolerance develops to the sympathomimetic but not the local anesthetic effects of cocaine. J Toxicol Clin Toxicol. 2000;38:719–27.
4. Stewart DJ, Inaba T, Lucassen M, Kalow W. Cocaine metabolism: cocaine and norcocaine hydrolysis by liver and serum esterases. Clin Pharmacol Ther. 1979;25:464–8.
5. Zhang JY, Foltz RL. Cocaine metabolism in man: identification of four previously unreported cocaine metabolites in human urine. J Anal Toxicol. 1990;14:201–5.
6. Warner A, Norman AB. Mechanism pf cocain hydrolysis and metabolism in vitro and in vivo: a clarification. Ther Drug Monit. 2000;22:266–70.
7. Matsubara K, Kagawa M, Fukui Y. In vivo and in vitro studies on cocaine metabolism: ecgonine methyl ester as a major metabolite of cocaine. Forensic Sci Int. 1984;26:169–80.
8. Cone EJ, Tsadik A, Oyler J, Darwin WD. Cocaine metabolism and urinary excretion after different routes of administration. Ther Drug Monit. 1098;20:556–60.
9. Harris DS, Everhart ET, Mendelson J, Jones RT. The pharmacology of cocaethylene in humans following cocaine and ethanol administration. Drug Alcohol Depend. 2003;72:169–82.
10. Wilson LD, Henning RJ, Suttheimer C, Lavins E, Balray E, Earl S. Cocaetylene cause dose dependent reductions in cardiac function in anesthetized dogs. J Cardiovasc Pharmacol. 1995;26:965–73.
11. Xu Y, Crumb WJ, Clarkson CW. Cocaethylene, a metabolite of cocaine and ethanol, is a potent blocker of cardiac sodium channels. J Pharmacol Expt Ther. 1994;271:319–25.
12. Cami J, Farré M, González ML, Segura J, de la Torre R. Cocaine metabolism in humans after use of alcohol. Clinical and research implications. Recent Dev Alcohol. 1998;14:437–55.
13. Clapp L, Martin B, Beresford TP. Sublingual cocaine – Novel recurrence of an ancient practice. Clin Neuropharmacol. 2004;27:93–4.
14. Nutt D, King LA, Saulsbury W, Blakemore C. Development of a rational scale to assess the harm of drugs of potential misuse. Lancet. 2007;369:1047–53.
15. Goodman LS, Gilman A, Brunton LL, Lazo JS, Parker KL. Goodman & Gilman's the pharmacological basis of therapeutics. New York: McGraw-Hill; 2006.
16. Abbott A. Neurobiological perspectives on drugs of abuse. TIPS. 1992;13:169.
17. Ramsey NF, van Ree JM. Reward and abuse of opiates. Pharmacol Toxicol. 1992;71(2):81–94.
18. Jaffe JH, Martin WR. Opioid Analgesics and Antagonists. In: Gilman AG, Rall TW, Nies AS, Taylor P, editors. The pharmacological basis of therapeutics. New York: McGraw-Hill; 1993. pp. 485–531.
19. Gay GR. Clinical management of acute and chronic cocaine poisoning. Ann Emerg Med. 1982;11:562–72.

20. Rapport RT, Gay GR, Inaba D. Propanolol: a specific antagonist to cocaine. Clin Toxicol. 1977;10:265–71.
21. Wetli CV, Fishbain DA. Cocaine-induced psychosis and sudden death in recreational cocaine users. J Forensic Sci. 1985;30:873–80.
22. Henning RJ, Wilson LD. Cocaethylene is as cardiotoxic as cocaine but is less toxic than cocaine plus ethanol. Life Sci. 1996;59:615–27.
23. Fowler JS, Volkow ND, Logan J, McGregor RR, Wang GF, Wolf AP. [11C]cocaine phrmaciokinertics in human brain and heart. Synapse. 1992;12:228–35.
24. Cami J, Farré M, González ML, Segura J, de la Torre R. Cocaine metabolism in humans after use of alcohol. Clinical and research implications. Recent Dev Alcohol. 1998;14:437–55.
25. Landry MJ. An overview of cocethylene, an alcohol-derived psychoactive, cocaine metabolite. J Psychoactive Drugs. 1992;24:273–6.
26. Brody SL, Slovis CM, Wrenn KD. Cocaine-related medical problems: consecutive series of 233 patients. Am J Med. 1990;88:325–31.
27. Om A. Cardiovascular complications of cocaine. Am J Med Sci. 1992;303:333–9.
28. Levis JT, Garmel GM. Cocaine-associated chest pain. Emerg Med Clin North Am. 2005;23: 1083–103.
29. Hollander JE, Henry TD. Evaluation and management of the patient who has cocaine-associated chest pain. Cardiol Clin. 2006;24:103–14.
30. Eichhorn EJ, Peacock E, Grayburn PA, et al. Chronic cocaine abuse is associated with accelerated atherosclerosis in human coronary arteries. J Am Coll Cardiol. 1992;19:105A.
31. Mouhaffel AH, Madu EC, Satmary WA, Fraker TD. Cardiovascular complications of cocaine. Chest. 1995;107:1426–34.
32. Levis JT, Garmel GM. Cocaine-associated chest pain. Emerg Med Clin North Am. 2005;23: 1083–103.
33. Hollander JE, Hoffman RS, Gennis P, et al. Prospective multicenter evaluation of cocaine-associated chest pain. Acad Emerg Med. 1994;1:330–9.
34. Anderson JL, Adams CD, Antman EM, et al. ACC/AHA 2007 guidelines for the management of patients with unstable angina/non-ST-elevation myocardial infarction. J Am Coll Cardiol. 2007;50:e1–e157.
35. Brogan WC, Lange RA, Kim AS, et al. Alleviation of cocaine-induced coronary vasoconstriction by nitroglycerin. J Am Coll Cardiol. 1991;18:581–6.
36. Hollander JE, Hoffman RS, Gennis P, et al. Nitroglycerin in the treatment of cocaine associated chest pain–clinical safety and efficacy. J Toxicol Clin Toxicol. 1994;32:243–56.
37. Hollander JE. The management of cocaine-associated myocardial ischemia. N Engl J Med. 1995;333:1267–72.
38. Honderick T, Williams D, Seaberg D, Wears R. A prospective, randomized, controlled trial of benzodiazepines and nitroglycerine or nitroglycerine alone in the treatment of cocaine-associated acute coronary syndromes. Am J Emerg Med. 2003;21:39–42.
39. Baumann BM, Perrone J, Hornig SE, et al. Randomized, double-blind, placebo-controlled trial of diazepam, nitroglycerin, or both for treatment of patients with potential cocaine-associated acute coronary syndromes. Acad Emerg Med. 2000;7:878–85.
40. Lange RA, Cigarroa RG, Flores ED, McBride W, Kim AS, Wells PJ, et al. Potentiation of cocaine-induced coronary vasoconstriction by beta-adrenergic blockade. Arch Intern Med. 1990;112:897–903.
41. Hollander JE, Henry TD. Evaluation and management of the patient who has cocaine-associated chest pain. Cardiol Clin. 2006;24:103–14.
42. Hollander JE, Carter WA, Hoffman RS. Use of phentolamine for cocaine-induced myocardial ischemia. N Engl J Med. 1992;327:361.
43. Hollander JE, Hoffman RS, Gennis P, Fairweather P, Feldman JA, Fish SS, et al. Cocaine-associated chest pain: one year follow-up. Acad Emerg Med. 1995;2:179–84.
44. Hollander JE, Thode HC, Hoffman RS. Chest discomfort, cocaine and tobacco. Acad Emerg Med. 1995;2:238.

45. Welch RD, Todd K, Krause GS. Incidence of cocaineassociated rhabdomyolysis. Ann Emerg Med. 1991;20:154–7.
46. Green R, Kelly KM, Gabrielson T, Levine SR, Vanderzant C. Multiple intracerebral hemorrhages after smoking "crack" cocaine. Stroke. 1990;21:957–62.
47. Levine SR, Brust JCM, Futrell N, Ho KL, Blake D, Millikan CH, et al. Cerebrovascular complications of the use of the "crack" form of alkaloidal cocaine. N Engl J Med. 1990;323: 699–704.
48. Westover AN, McBride S, Haley R. Stroke in young adults who abuse amphetamines or cocaine. A population-based study of hospitalized patients. Arch Gen Psychiatry. 2007;64:495–592.
49. Schindler CW. Cocaine and cardiovascular toxicity. Addict Biol. 1996;1:31–47.
50. Karch SB, Billingham ME. The pathology and etiology of cocaine-induced heart disease. Arch Pathol Lab Med. 1988;112:225–30.
51. Lange RA, Cigarro RG, Yancy CWJ, Willard JE, Pompa JJ, Sills MN, et al. Cocaine induced coronary artery vasoconstriction. N Eng J Med. 1989;321:1557–62.
52. Robertson MW, Leslie CA, Bennett JPJ. Apparent synaptic dopamine deficiency induced by withdrawal from chronic cocaine treatment. Brain Res. 1991;538:337–9.
53. Neugebauer V, Zinebi F, Russell R, Gallagher JP, Shinnick-Gallagher P. Cocaine and kindling alter the sensitivity of group II and III metabotropic glutamate receptors in the central amygdala. J Neurophysiol. 2000;84:759–70.
54. Marley RJ, Gioldberg SR. Pharmacogenetic assessment of the effects of carbamazepine on cocaine-kindled and cocaine-induced seizures. Brain Res. 1992;579:43–9.
55. Iccocioppo R, Sanna PP, Weiss F. Cocaine predictive stimulus induces drug seeking behavioral and neural actions in limbic brain regions after multiple months of abstinence. Proc Natl Acad Sci USA. 2001;98:1976–81.
56. Lidow MS, Song ZM. Primates exposed to cocaine in utero display reduced density and number of cerebral cortical neurons. J Comp Neurol. 2001;435:263–75.
57. Lidow MS, Song ZM. Effect of cocaine on cell proliferation in the cerebral wall of monkey fetuses. Cereb Cortex. 2001;11:545–51.
58. Kosofsky BE, Hyman SE. No time for complacency: the fetal brain on drugs. J Comp Neurol. 2001;435:259–62.
59. Smith LM, Qwesbi N, Renslo R, Sinow RM. Prenatal cocaine exposure and cranial sonographic findings in preterm infants. J Clin Ultrasound. 2001;29:72–7.
60. Frank DA, Augustyn M, Knight WG, Pell T, Zuckerman B. Growth, development, and behavior in early childhood following prenatal cocaine exposure: a systematic review. JAMA. 2001;285:1613–25.
61. Rawson RA, Mc Cann MJ, Hasson AJ, Ling W. Addiction pharmacotherapy 2000: new options, new challenges. J Psychoactive Drugs. 2000;321:371–8.
62. Carrera MRA, Ashley JA, Wirsching P, Koob GF, Janda KD. A second generation vaccine protects against the psychoactive effects of cocaine. Proc Natl Acad Sci USA. 2001;98:1988–91.
63. Matsushita M, Hoffman TZ, Ashley JA, Zhou B, Wirsching P, Janda KD. Cocaine catalytic antibodies: the primary importance of linker effects. Bioorg Medi Chem Lett. 2001;11:87–90.
64. Hayashi T, Su TP. Regulating ankyrin dynamics: roles of sigma 1 receptors. Proc Natl Acad Sci USA. 2001;98:491–6.
65. Schneider-Helmert D, Spinweber CL. Evaluation of L-tryptophan for treatment of insomnia: a review. Psychopharmacology. 1986;89:1–7.
66. Hartmann E. Effects of L-tryptophan on sleepiness and on sleep. J Psychiatr Res. 1982;17: 107–13.
67. Lam RW, Levitan RD, Tam EM, Yatham LN, Lamoureux S, Zis AP. L-tryptophan augmentation of light therapy in patients with seasonal affective disorder. Can J Psychiatr. 1997;42:302–6.
68. Levitan RD, Shen JH, Jindal R, Driver HS, Kennedy SH, Shapiro CM. Preliminary randomized double-blind placebo-controlled trial of tryptophan combined with fluoxetine to treat major depressive disorder: antidepressant and hypnotic effects. J Psychiatr Nuerosci. 2000;25:337–46.

69. Meyers S. Use of neurotransmitter precursors for treatment of depression. Alternat Med Rev. 2000;5:64–71.
70. Turner EH, Loftis JM, Blackwell AD. Serotonin a la carte: supplementation with the serotonin precursor 5-hydroxytryptophan. Pharmacol Ther. 2006;109:325–38.
71. Chinevere TD, Sawyer RD, Creer AR, Conlee RK, Parcell AC. Effects of L-tyrosine and carbohydrate ingestion on endurance exercise performance. J Appl Physiol. 2002;93:1590–7.
72. StrŸder HK, Hollmann W, Platen P, Donike M, Gotzmann A, Weber K. Influence of paroxetine, branched-chain amino acids and tyrosine on neuroendocrine system responses and fatigue in humans. Horm Metab Res. 1998;30:188–94.
73. Thomas JR, Lockwood PA, Singh A, Deuster PA. Tyrosine improves working memory in a multitasking environment. Biochem Behav. 1999;64:495–500.
74. Deijen JB, Orlebeke JF. Effect of tyrosine on cognitive function and blood pressure under stress. Brain Res Bull. 1994;33:319–23.
75. Oliveto AH, Feingold A, Schottenfeld R, Jatlow P, Kosten TR. Desipramine in opioid-dependent cocaine abusers maintained on buprenorphine vs methadone. Arch Gen Psychiatry. 1999;56:812–20.
76. Sherer MA, Kumor KM, Jaffe JH. Effects of intravenous cocaine are partially attenuated by haloperidol. Psychiatry Res. 1989;27:117–35.
77. Schneier FR, Siris SG. A review of psychoactive substance use and abuse in schizophrenia: patterns of drug choice. J Ner Ment Dis. 1987;175:641–52.
78. Tennant FS, Sagherian AA. Double blind comparison of amantadine and bromocriptine for ambulatory withdrawal from cocaine dependence. Arch Intern Med. 1987;147:109–12.
79. Grabowski J, Schmitz J, Roache JD, Rhoades H, Elk R, Creson DL, editors. Methylphenidate (MP) for initial treatment of cocaine dependence and a model for medication evaluation. Problems of Drug Dependence Proceedings of the 55th Annual Scientific Meeting, The College on Problems of Drug Dependence, Inc; 1993; Washington, DC: NIH Pub. No. 94–3749. Supt. of Docs., US Govt. Print. Off. 1994.
80. Brodie JD, Figueroa E, Laska EM, Dewey SL. Safety and efficacy of γ-vinyl GABA (GVG) for the treatment of methamphetamine and/or cocaine addiction. Synapse. 2005;55:122–5.
81. Carroll KM, Fenton LR, Ball SA, Nich C, Frankforter TL, Shi J, et al. Efficacy of disulfiram and cognitive behavior therapy in cocaine-dependent outpatients. A randomized placebo-controlled trial. Arch Gen Psychiatry. 2004;61:264–72.
82. Romach MK, Glue P, Kampman K, et al. Attenuation of the euphoric effects of cocaine by the dopamine D1/D5 antagonist ecopipam (SCH 39166). Arch Gen Psychiatry. 1999;56:1101–6.
83. Marx J. How stimulant drugs may calm hyperactivity. Science. 1999;283:306–9.
84. Conrad KL, Tseng KY, Uejima KL, Reimers JM, Heng L-J, Shaham Y, et al. Formation of accumbens GluR2-lacking AMPA receptors mediates incubation of cocaine craving. Nature. 2008;454:118–21.

Part II
Methamphetamine-an Old Aquaintance in New Clothes

History of Methamphetamine

Amphetamine, phenylisopropylamin or alpha-methylphenyl-ethylamin was originally isolated from the plant Ephedra vulgaris by the chemist Sedalano in 1887, and was later, synthesized from ephedrine in Japan by the chemist Nagayoshi Nagai in 1893 [2]. It was not until 1889, when the pharmaceutical company Merck marketed it as a remedy against cold. Under the nomenclature system of that time, amphetamine was originally called alpha-methylphenylethylamine, which was shortened to "amphetamine". Later in 1930 it was discovered that amphetamine raised blood pressure. In 1932, after it was found to shrink mucous membranes, the pharmaceutical company Smith, Kline and French started to market it as a nasal inhaler containing amphetamine under the trade name Benzedrine™, a racemic mixture of dextro- and levo-amphetamine (dl-amphetamine) to treat nasal congestion (Fig. 3). In addition, from 1938 on the pharmaceutical company Boehringer/Ingelheim from Germany also marketed Benzedrine™ until it became a prescription drug in 1970. Early users of the Benzedrine™ inhaler discovered that it had a euphoric stimulant effect, which resulted in becoming one of the earliest synthetic stimulants widely used for recreational (i.e., non-medical) purposes. Even though this drug was intended for inhalation, many people abused it by cracking the container open and putting a paper strip inside, which soaked the Benzedrine™. The strips were often rolled into small balls and swallowed, or taken with coffee or alcohol. The drug was often referred to as "Bennies" by users and in the literature. Because of the stimulant side effect, physicians discovered that amphetamine could also be used to treat narcolepsy. This led to the production of benzedrine in tablet form, and often doctors who used to perk up lethargic patients, also used benzedrine before breakfast to get them though the day. In the 1940s and 1950s reports began to emerge about the abuse of Benzedrine™ inhalers, and in 1949, doctors began to move away from prescribing benzedrine as a bronchodilator and appetite suppressant. In 1959, the FDA made it a prescription drug in the United States.

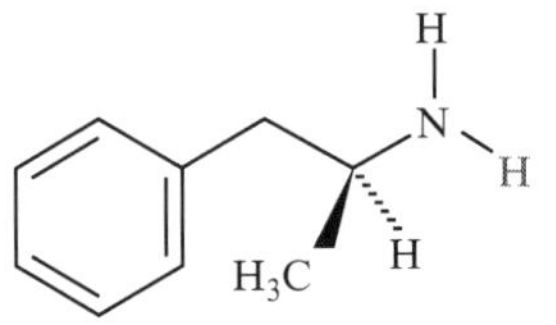

(S)-(+)-amphetamine or Dexedrine
The blue H is replaced by CH₃ in methamphetamine

Fig. 3 The dextroratatory isomer of amphetamine, and its conversion to D-methamphetamine, the pharmacologically active ligand. During synthesis, it yields the racemic mixture, containing the dextro- and the levo-isomer

From Amphetamine to Methamphetamine

As early as 1919, Akira Ogata synthesized methamphetamine via reduction of ephedrine using red phosphorus and iodine. Later, the chemists Hauschild, and Dobke from the pharmaceutical company Temmler in Germany developed an easier method for converting ephedrine to methamphetamine. Now it was possible to market it on a large scale under the trade name Pervitin® (1-phenyl-2-methylaminopropanhydrochloride) as a non-prescription drug (Fig. 4). It was not until 1986 that Pervitin® became a controlled substance agent needing a special prescription.

Although already being used as a stimulant and a drug to give stamina and endurance in the years between both world wars, the widespread and frequent uses of methamphetamine was during World War II when the German military dispensed it under the trade name Pervitin® [3]. It was widely distributed across rank and division, from elite forces to tank crews and aircraft personnel. Chocolates dosed with methamphetamine were known as *Fliegerschokolade* or *Stuka-Pillen* ("flyer's chocolate" or "Dive bomber pills") when given to pilots, or *Panzerschokolade* ("tanker's chocolate") when given to tank crews [4]. The latter stimulants, due to their side effects, have very little medical benefits, but have nevertheless always found a use in warfare to avoid fatigue and to take away soldiers' inhibitions when faced with dangerous missions. Also, from 1942 until his death in 1945, the personal physician of Adolf Hitler, Theodor Morell gave frequent intravenous injections of methamphetamine as a treatment for depression and fatigue. It is likely that it was also used to treat Hitler's speculated Parkinson's disease, since he had daily injections of methamphetamine during the last 3 years of the war, or that his Parkinson-like symptoms, which developed from 1940 onwards, were related to use of methamphetamine [5]. In addition to the use of the pure agent, experiments were performed in volunteers in Kiel/Germany where methamphetamine 3 mg was combined with cocaine 5 mg and eukodal 5 mg (a morphine derivative) in order to further increase endurance, fight fatigue and produce stamina in one-manned torpedo submarine fighters during the last days of World War II, under the code name P-IX-Pill [6].

Fig. 4 Label for Pervitin® tablets (=1-Phenyl-2-methylamino-propan-hydrochloride = methamphetamine) as it was used in the 1920–1930 era, produced by the German pharmaceutical company Temmler until 1988

Fig. 5 Amphetamines were a regular additive not only for the aviators to increase alertness and fight fatigue, but also they were (and they still are) a regular ingredient to boost up the fighting moral in soldiers in all armies in World War II

Because of its stimulating effects, American, British, German and Japanese forces used amphetamines widely. Bomber crews used amphetamines to maintain alertness on long night missions; watch-keepers on ships used them to stay alert too (Fig. 5). But amphetamines are still in use as pep pills today. On April 17, 2002, USAF bomber pilots on amphetamine "*go pills*" mistakenly bombed Canadian forces in Afghanistan, killing four and wounding eight. The USAAF says that the amphetamine was issued to keep them awake on long missions just like 60 years ago.

Although being known for its effect resulting in an increase in alertness and dampening of sleep, the first really large-scale methamphetamine abuse occurred in Japan immediately after World War II, when stocks formerly used by the military (and supplied by Germany) flooded the market, and it took a strict campaign to bring things under control. After World War II, a large supply of amphetamine, formerly stockpiled by the Japanese military, became available in Japan under the street name "*shabu*" also Philopon, or Hiropon, as a trade name [7]. The Japanese Ministry of Health banned it in 1951, and its prohibition is thought to have added to the growing yakuza-activities related to illicit drug production. Today, methamphetamine is still associated with the Japanese underworld, but its usage is discouraged by strong social taboos.

In the Western world, the 1950s saw a rise in the legal prescription of methamphetamine to the American and the European public. According to the 1951 edition of Pharmacology and Therapeutics by Arthur Grollman, it was prescribed for "*narcolepsy, post-encephalitic Parkinsonism, alcoholism, ... in certain depressive states ... and in the treatment of obesity.*" The 1960s saw the start of significant use of clandestinely manufactured methamphetamine as well as methamphetamine created in users' own homes for personal use. The recreational use of methamphetamine peaked in the 1980s, San Diego, becoming "*the methamphetamine capital of North America*". After 1951, amphetamines became prescription drugs in the USA and were widely prescribed as pick-me-ups. They were once used as a slimming aid, as they reduced appetite, but in large scale amphetamine tablets became known as "*pep pills*" since 1950s. Long distance lorry drivers used amphetamines to help them stay awake, and students knew that amphetamines helped them stay alert when studying, and helped them score better marks. Under the influence of amphetamines, the poet Jack Kerouac produced "*On the Road*" in one continued frenzy of typing in 3 weeks in 1951.

In the UK, the 1964 Drugs Prevention of Misuse Act made unlicensed possession and import of amphetamines illegal. But they continued to be widely prescribed, both in the UK and the USA. Illegal production in the USA has been associated with biker gangs as it has spread from San Francisco into the Midwest, while amphetamines powered much of 1960s British youth culture. For many years, methamphetamine was "available" in tablet form or as a powder. In the late 1980s, "crystal meth" or "ice" became available in Hawaii, produced in Far Eastern laboratories by careful recrystallisation of methamphetamine hydrochloride, resulting in quite large, clear, crystals (Fig. 6). Though a salt, and unlike amphetamine sulphate, it is volatile enough to vaporize unchanged resulting in a spread to the American mainland. Methamphetamine abuse has become a real problem right across the country, becoming the leading illicit drug in rural USA, while in Japan and Southeast Asia

Fig. 6 Methamphetamine crystals, where only the higher purity is related to the word *"Ice"*

it has become endemic, controlled by yakuza gangs, including countries like Cambodia and Thailand, where is known as "yaa baa" (yaa = medicine, baa = crazy, after its propensity to make people behave in crazy, uncontrolled ways.

Methamphetamine started to be known as "speed" in the early 1960s, possibly as a consequence of its use in "speedballs" with heroin, and also because of its go-faster characteristics. In the Haight-Ashbury region of San Francisco, the *"Summer of Love"* in 1967 turned sour the following year when many residents had moved on from marijuana and psychedelics like LSD to injecting methamphetamine intravenously. The resulting epidemic of methamphetamine abuse and consequent illnesses (and crime) led to a change in the neighborhood to a less friendly, indeed dangerous, one, and the once famous motto "peace and love" changed to "speed kills". Among the many celebrated victims of methamphetamine abuse, Johnny Cash, the American singer, survived, as he found God. Others were not so lucky, as methamphetamine contributed to the deaths of Judy Garland and Edie Sedgwick, Andy Warhol's muse.

During Vietnam both the Air Force and Navy made methamphetamine available to aviators. Intermittently since Vietnam up through Desert Storm the Air Force has used both methamphetamine and sedatives in selected aircraft for specific missions. Following Desert Storm an anonymous survey of deployed fighter pilots was completed. Surveys (464) were returned (43%). For Desert Storm 57% used stimulants at some time; 17% routinely, 58% occasionally, and 25% only once. Within individual units, usage varied from 3% to 96%, with higher usage in units tasked for sustained combat patrol (CAP) missions. Sixty one percent of those who used stimulants reported them essential to mission accomplishment.

> *Presently, possession of amphetamine/methamphetamine without a prescription, is illegal in the USA under the Controlled Substances Act of 1970, as they are Schedule II drugs.*

Mode of Action of Methamphetamine

When methamphetamine is first used, the effects, aside from a simple stimulant, include a sense of euphoria, openness, and intellectual expansion. There can be a mild psychedelic component as the new user feels they see the world a little differently. The sensation of "mind-expansion" and openness quickly fade after the first few uses and physical and mental stimulation dominate the observed effects.

Methamphetamine also causes increased heart rate and blood pressure and can cause irreversible damage to blood vessels in the brain, producing strokes. Other effects of methamphetamine include respiratory problems, irregular heartbeat, and extreme anorexia. Abusers of methamphetamine are prone to accidents because the drug produces excitation and grandiosity followed by excess fatigue and sleeplessness. Taken intravenously, methamphetamine may lead to serious antisocial behavior and can precipitate a schizophrenic episode. Cell death of dopaminergic neurons, however, has not been documented in methamphetamine abusers, which could explain why extended abstinence allows for some recovery from methamphetamine-induced deficits in dopamine function (Fig. 7). But even though a recent neuroimaging study of methamphetamine abusers showed partial recovery of brain function in some regions following protracted abstinence, function in other regions did not display recovery even after 2 years of abstinence, suggesting that long-lasting and even permanent brain changes may result from methamphetamine abuse.

In contrast to crack cocaine, the effects of methamphetamine are intense and persist longer than the brief "high" of cocaine. Immediately after smoking or following intravenous injection, the methamphetamine user experiences an intense sensation, called a "rush" or "flash," that lasts only a few minutes and is described as extremely pleasurable. Oral or intranasal use produces euphoria – a high, but not a rush. Users may become addicted quickly, and use it with increasing frequency and in increasing doses. And although methamphetamine induces a tolerance that develops slowly, there is the need to progressively increase the dose, until finally the amounts are several hundredfold greater than the original therapeutic dose that eventually was ingested or injected.

Tolerance to various effects develops unequally. Tachycardia and enhanced alertness are diminished but psychotoxic effects, such as hallucinations and delusions, still occur. Contrary to cocaine, even massive doses are rarely fatal. Long-term users

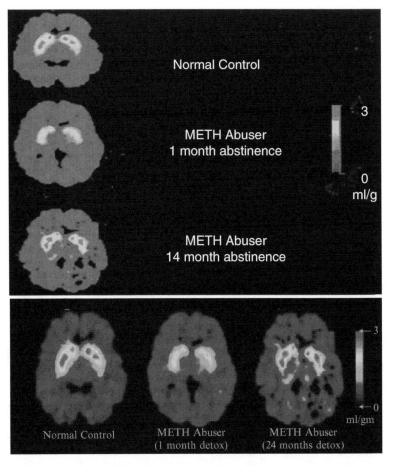

Fig. 7 PET in a chronic methamphetamine user showing reduced activity within the dopaminergic basal ganglia, and its slow recovery after abstinence. Partial recovery of brain dopamine transporters in methamphetamine (METH) abuser after protracted abstinence (Adapted from [8])

have reportedly injected as much as 15,000 mg of amphetamine in 24 h without observable acute illness. It has been proposed that sensitization to cocaine and amphetamines results from increased synaptic connectivity due to repeated exposure to the drugs. Circumstances surrounding drug taking (environmental cues) can trigger this neuronal hypersensitivity and exert a powerful influence leading to physiological arousal and increased craving at the mere sight or thought of the drug [9] to a degree, the brain may become "rewired" for stimulant addiction.

Interaction of Methamphetamine with other Stimulants

Some psychostimulants are known to have paradoxical effects in the methamphetamine user, such as methylphenidate (Ritalin®), which exerts a calming effect in children with ADH (attention-deficit hyperactivity). New research in animals suggests that these stimulants act by facilitating therapeutic concentrations of serotonin, rather than affecting dopamine [10].

Pharmacology of Methamphetamine

Methamphetamine increases dopamine levels in the central nervous system (CNS) by as much as 2,600%, primarily by stimulating presynaptic release of the neurotransmitter, rather than by re-uptake blockade [9, 11]. Such increase has the effect of stimulating regions of the brain linked with vigilance and the action of the heart [11]. For a short while, the user feels sharper, stronger and more energetic. It is believed that a significant proportion of the dopamine-producing cells in the brain can be damaged by prolonged exposure to even low levels of methamphetamine, and this is responsible for reduced levels of dopamine; this can affect memory-, attention- and decision-taking functions. Chronic methamphetamine abuse is reported to lead to significant reduction in grey matter in the brain, greater than those in dementia or schizophrenia patients, though this needs further clarification. Associated health risks involve social and family problems, including risky sexual behavior. Drug-induced psychosis may result. Methamphetamine effects last for days in the body, and some degree of neurological impairment may last up to 2 years or more after cessation of the drug [12].

Thus, methamphetamine is a potent central nervous system stimulant which affects neurochemical mechanisms responsible for regulating heart rate, body temperature, blood pressure, appetite, attention, mood and responses associated with alertness or alarm conditions. The acute effects of the drug closely resemble the physiological and psychological effects of an epinephrine-provoked fight-or-flight response, including increased heart rate and blood pressure, vasoconstriction, bronchodilation, and hyperglycemia (Fig. 8). Users experience an increase in focus, increased mental alertness, and the elimination of fatigue, as well as a decrease in appetite (Table 1). The methyl group in methamphetamine is responsible for the potentiation of effects as compared to the related compound amphetamine, rendering the substance on the one hand more lipid soluble and easing transport across the blood brain barrier, and on the other hand more stable against enzymatic degradation by monamine oxidase inhibitor (MAOI). Methamphetamine causes the norepinephrine, dopamine (Fig. 9) and serotonin (5-HT) transporters to reverse their direction of flow. This inversion leads to a release of these transmitters from the vesicles to the cytoplasm and from the cytoplasm to the synapse (releasing monoamines in rats with ratios of about NE:DA = 1:2, NE:5-HT = 1:60), causing increased stimulation of post-synaptic receptors. Methamphetamine also indirectly

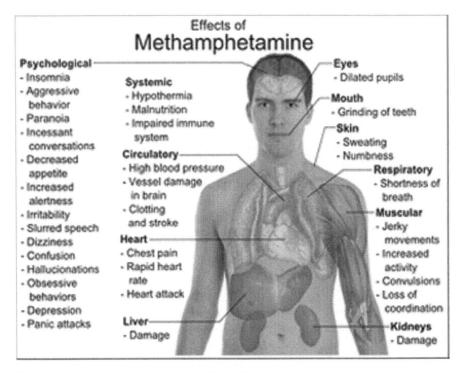

Fig. 8 Summary of the pharmacological effects of methamphetamine

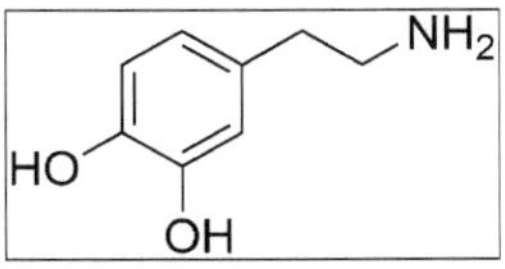

Fig. 9 Molecular structure of dopamine, an important neurotransmitter abundantly present in the brains mesolimbic *reward center* after methamphetamine

prevents the re-uptake of these neurotransmitters, causing them to remain in the synaptic cleft for a prolonged period (inhibiting monoamine re-uptake in rats with ratios of about: NE:DA = 1:2.35, NE:5-HT = 1:44.5 [13].

Methamphetamine is also a potent neurotoxin, shown to cause dopaminergic degeneration [14, 15], with long-lasting damage to dopamine nerve endings in the striatum [14], and high doses of methamphetamine produce loss in several markers of brain dopamine and serotonin neurons [15]. Since dopamine and serotonin concentrations, dopamine and 5-HT uptake sites, and tyrosine and tryptophan hydroxylase activities are reduced after the administration of methamphetamine [16] it has been assumed that dopamine plays a role in methamphetamine-induced neurotoxicity. This assumption was underlined by experiments, which reduce dopamine production or block the release of dopamine decrease the toxic effects of methamphetamine administration. Since dopamine breakdown produces reactive oxygen species (ROS) such as hydrogen peroxide, it is likely that the oxidative stress that occurs after taking methamphetamine mediates its neurotoxicity [17], as it has been

Table 1 Summary of the different effects induced by methamphetamine

Positive effects
- Increased energy and alertness
- Decreased need for sleep
- Euphoria
- Increased sexuality
- Decreased need for food

Neutral effects
- Excessive talking
- Weight loss
- Sweating
- Visual and auditory hallucinations (hearing voices)

Negative effects
- Disturbed sleep patterns
- Tightened jaw muscles, grinding teeth (trismus and bruxia)
- Loss of appetite (anorexia), leading to poor nutrition and weight loss with heavy use
- Reduced enjoyment of eating
- Loss of interest in sex, over time
- Itching, welts on skin
- Nausea, vomiting, diarrhea
- Excessive excitation, hyperactivity
- Shortness of breath
- Moodiness and irritability
- Anxiousness and nervousness
- Aggressiveness
- Panic, suspiciousness and paranoia
- Involuntary body movements (uncontrollable movement and/or twitches of fingers, facial and body muscles, lip-smacking, tongue protrusion, grimacing, etc.)
- False sense of confidence and power (delusions of grandeur)
- Aggressive and violent behavior
- Severe depression, suicidal tendencies

Effects with habitual use
- Fatal kidney and lung disorders
- Possible brain damage
- Permanent psychological problems
- Lowered resistance to illnesses
- Liver damage
- Stroke
- Deep or disturbed sleep lasting up to 48 h
- Extreme hunger
- Psychotic reaction
- Anxiety reactions

demonstrated that a high ambient temperature increases the neurotoxic effects of methamphetamine [18]. Also microglia, the major antigen-presenting cells in brain and when activated, secrete an array of factors that cause neuronal damage may be involved in neurotoxicity. Surprisingly, very little work has been directed at the study of microglial activation as part of the methamphetamine neurotoxic cascade [19].

In addition recent research indicates that methamphetamine binds to a group of receptors called trace amine-associated receptors (TAARs) [20]. This newly discovered receptor system seems to be affected by a range of amphetamine-like substances, which have shown to mediate the identification of social cues in mice, and are also present in humans and fish. TAARs are G-protein coupled receptors. TAARs are not, however, related to odourant receptors. Analysis of the DNA sequence showed that they are most closely related to receptors for the neurotransmitters dopamine and 5-hydroxytryptamine (5-HT). One of the TAAR receptors binds to beta-phenylethylamine, an amine compound whose concentration in mouse urine is elevated in response to stress. At least three bind to isoamylamine, a pheromone found in the urine of male, but not female, mice [21]. To a lesser extent methamphetamine acts as a dopaminergic and adrenergic re-uptake inhibitor and in high concentrations as a monamine oxidase inhibitor (MAOI). Since it stimulates the mesolimbic reward pathway, causing euphoria and excitement, it is prone to abuse and addiction. Users may become obsessed or perform repetitive tasks such as cleaning, hand washing, or assembling and disassembling objects. Withdrawal is characterized by excessive sleeping, eating and depression-like symptoms, often accompanied by anxiety and drug-craving [22]. Users of methamphetamine sometimes take sedatives such as benzodiazepines as a means of easing their "*come down*".

Prescribing Methamphetamine

The dextrorotatory isomer of methamphetamine, i.e., dextromethamphetamine marketed under the brand name Desoxyn® (Fig. 10), can be prescribed to treat attention-deficit hyperactivity (ADH) disorder, though unmethylated amphetamine is more commonly prescribed. It is also considered a second line of treatment in narcolepsy and obesity, when amphetamine and methylphenidate cause the patient too many side effects. It is only recommended for short term use (~6 weeks) in obesity patients because it is thought that the anorectic effects of the drug are short lived and produce tolerance quickly, whereas the effects on CNS stimulation are much less susceptable to tolerance. It is also used illegally for weight loss and to maintain alertness, focus, motivation, and mental clarity for extended periods of time, and for recreational purposes.

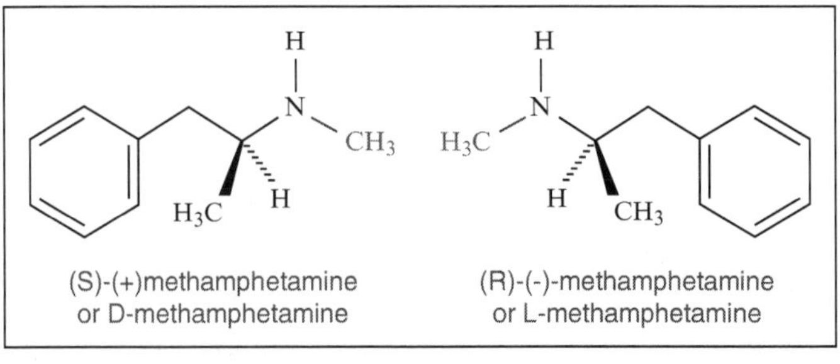

Fig. 10 Molecular structure of both optical isomers of methamphetamine differing only in their ability to rotate plane-polarized light

Chemistry of Methamphetamine

Methamphetamine contains a chiral carbon atom, and therefore exists as two optical isomers, non-superimposable mirror images of each other (Fig. 10). The isomers have identical chemical reactions, solubilities and melting points. It is possible that they have different smells, as the isomers of amphetamine itself are reported to differ. They just differ in their opposite rotations of plane-polarized light and, because they have different 3-D structures, they fit chiral protein receptors differently, and therefore have different effects upon the body.

Laboratories in many places, notably throughout the USA, are engaged in clandestine manufacture of methamphetamine. The chemistry is well understood, and widely published, notably on the web, though not easy to achieve. It is the synthesis of methamphetamine that is illegal. Two methods that have been described both involve the reduction of ephedrine (Figs. 11, 12), which by itself is readily available,

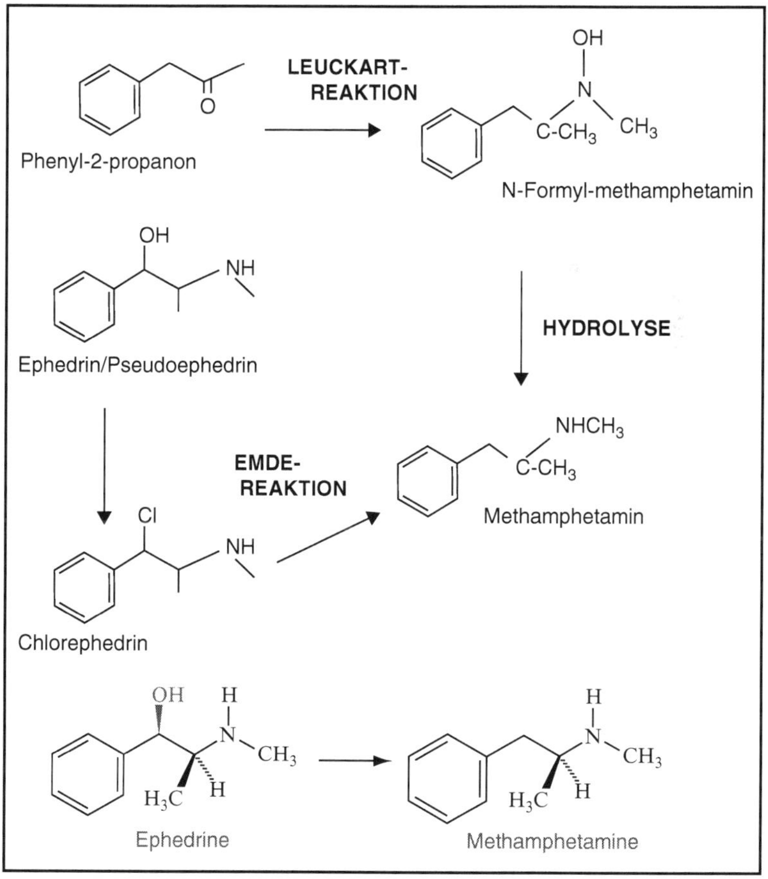

Fig. 11 Different chemical pathways for the synthesis of methamphetamine manufacturing from ephedrine

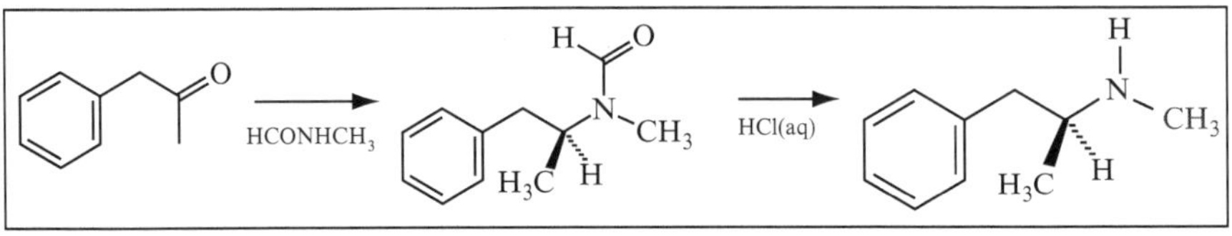

Fig. 12 Phenylacetone reaction to formyl amide of methamphetamine, which eventually results in the formation of D-methamphetamine

Table 2 Summary of physicochemical properties of methamphetamine

Appearance	Colorless, odorless, crystal-like powder; pure methamphetamine has an oil-like texture
Molecular weight	129,4
Chemical formula of base	$C_{10}H_{15}N$
Melting point	Methamphetamine hydrochloride 173°C
Solubility of pure methamphetamine	Good soluble in ether, ethanol, insoluble in water
Solubility of the salt	Soluble in water, ethanol, chloroform, insoluble in ether
Metabolism	Liver enzymes, glucuronidation, and N-demethylation
Mean distribution time = max onset of action	20–40 min
Man plasma half-time = mean duration of action	7–34 h
Therapeutic dose	3–9 mg

using either lithium in liquid ammonia as the reducing agent, or else red phosphorous and iodine to generate the hydrolized form for the reduction.

A further process involves the reductive amination of phenylacetone, first converting it into the formyl amide of methamphetamine, then hydrolyzing the amide with HCl.

Much of this synthetic chemistry is difficult and dangerous outside a proper laboratory, since fires and explosions, as well as chemical burns to individuals, are commonly reported. Because the free amphetamine and methamphetamine molecules are rather insoluble oily liquids, they are usually handled as salts, such as methamphetamine hydrochloride (Table 2). This salt dissociates in the body, liberating the free amine and causes the known sympathomimetic effects.

Crystal Methamphetamine

Crystal methamphetamine or "ice" typically resembles small fragments of glass or shiny blue-white "rocks" of various sizes (Fig. 13). Contrary to powdered methamphetamine (the other form of d-methamphetamine hydrochloride), which contains by-products from synthetic preparation, crystal methamphetamine has a higher purity level and produces longer-lasting and more intense euphoric effects than the powdered form of the drug.

Crystal methamphetamine typically is smoked using glass pipes similar to pipes used to smoke crack cocaine (Fig. 14). Crystal methamphetamine also may be injected. A user who smokes or injects the drug immediately experiences an intense sensation followed by a high that may last for 12 h or more.

Crystal methamphetamine is used by individuals of all ages and is increasingly gaining in popularity as a club drug in discotheques (Fig. 15). It is difficult to determine how many individuals in the United States or other Western countries use crystal methamphetamine because most illicit drug use surveys do not distinguish between crystal methamphetamine and powdered methamphetamine. Those surveys that do draw such a distinction reveal that use of crystal methamphetamine is prevalent. According to the University of Michigan's Monitoring the Future Survey, nearly 5% of high school seniors in the United States used crystal methamphetamine at least once in their lifetime and 3% used the drug in the past year.

Addictive Properties of Methamphetamine

As with other psychoactives, different routes of administration have different profiles of effects. Oral methamphetamine ingestion tends to lack rushing, has less euphoric effects, and tends to cause far less of a feeling of wanting to do it again than the other methods. Smoking and injecting methamphetamine are associated with stronger, more euphoric effects and these are more associated with compulsive/addictive user patterns. The most frequent route of administration among primary methamphetamine/ amphetamine substance abuse treatment admissions was smoking (63%) followed by injection (19%; Fig. 16). Thirteen percent of primary methamphetamine/amphetamine admissions reported inhalation and 5% reported oral consumption.

Fig. 13 Appearance of methamphetamine hydrochloride being a powder (*left*) and *Crystal-Meth*, which due to its higher purity has a crystal-like appearance being named *ICE* (*right*)

Fig. 14 Preferential abuse of crystal being smoked in glass pipes and/or snorted rather than taken as a pill

Cook [23] measured plasma levels of methamphetamine after oral administration and after smoking. After an oral dose of 0.25 mg/kg (17.5 mg for a 70 kg person), plasma levels began to rise 30 min after ingestion and reached peak levels (approximately 38 ng/mL) at about 3 h. Plateau levels were maintained for about 3–4 h and slowly

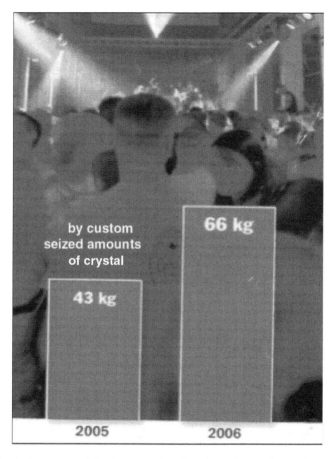

Fig. 15 Rise in the amount of the drug *crystal* methamphetamine confiscated by customs when being sold in discotheques over the past years in Germany

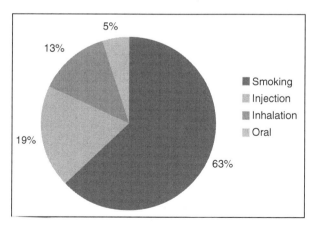

Fig. 16 Primary methamphetamine/amphetamine treatment admissions by route of administration (From SAMHSA Treatment Episode Data Set [TEDS] 2005)

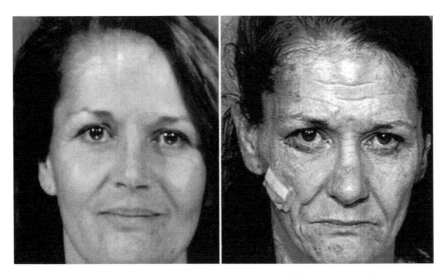

Fig. 17 Photos of a chronic consumer of methamphetamine before and after 3 years of illicit use

declined over the next 4 h. After smoking a roughly equivalent dose (about 21 mg/subject), methamphetamine plasma levels approximated 80% of peak levels within minutes and peaked at a similar level after oral administration (about 42 ng/mL) at about 2 h after administration. Plateau levels were maintained for another 2 h and then slowly declined over the next 4 h. By comparison, plasma levels of smoked cocaine (21–22 mg/subject) peaked at approximately 240 ng/mL at about 5–10 min after administration, and then declined rapidly, dropping to 50% of maximum levels within 1 h. In contrast to the 1 h half-life of smoked cocaine, the half-life of smoked methamphetamine was found to be 11–12 h. Thus, methamphetamine reaches a plateau and declines much more slowly than cocaine, which is rapidly eliminated. These considerations suggest the possibility of accumulation of methamphetamine (but not cocaine) in the body with repeated dosing [23].

Methamphetamine due to its potency on the release of norepinephrine, dopamine and serotonin is highly addictive [13], especially when injected or smoked. While not life-threatening, withdrawal is often intense and, as with all addictions, and inspite of Step meetings such as Crystal Meth Anonymous, relapse is common with a consequential decline of appearance with a decay of memory and emotions in the reward system (Figs. 17, 18).

Addictive Properties of Methamphetamine

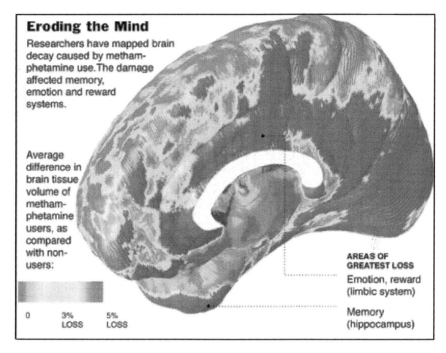

Fig. 18 PET image of the effects of methamphetamine on the CNS. (B courtesy of Dr. Phil Thompson from UCLA)

Treatment Options in Methamphetamine Addiction and Withdrawal

Primarily, therapy for a *"meth-addict"* includes the importance of cutting ties with friends who engage in substance abuse. Not doing so will make it more difficult for the patient to live without the temptation. Abrupt interruption of chronic methamphetamine use results in the withdrawal syndrome in almost 90% of the cases. Withdrawal of methamphetamine often causes a depression, which is even longer and deeper than the depression from cocaine withdrawal [12, 24]. Users may experience fatigue, long, disturbed periods of sleep, irritability, intense hunger, and moderate to severe depression. The length and severity of the depression is related to how much and how often amphetamines were used including the following symptoms [25]:

- Craving
- Exhaustion
- Depression
- Mental confusion
- Restlessness and insomnia
- Skin changes such as abscesses or boils (Fig. 17)
- Tooth wear due to lack of hygiene

Continued use of high doses of methamphetamine produces anxiety reactions during which the person is fearful, tremulous, and concerned about his physical well being, an amphetamine psychosis in which the person misinterprets others' actions, hallucinates, and becomes unrealistically suspicious, sometimes resulting in a violent behavior. After such stimulatory phase, an exhaustion syndrome, involving intense fatigue and the need for sleep, together with a prolonged depression (during which suicide is possible) is the result of a depletion of serotonin and dopamine in basal ganglia (Fig. 7). Especially users of large amount of amphetamines over a long period of time can develop such an amphetamine psychosis, a mental disorder, which is similar to paranoid schizophrenia. Symptoms usually disappear within a few weeks after drug use stops. Recovery from prolonged amphetamine psychosis is usual.

Thoroughly disorganized and paranoid users recover slowly but completely. The more florid symptoms fade within a few days or weeks, but some confusion, memory loss, and delusional ideas commonly persist for months.

Medical Treatment in Methamphetamine Related Depression

Medical treatments include the use of antidepressant agents such as imipramine, desipramine, amitriptyline, dosepin, trazodone, or fluoxetine (Prozac®). These agents affect serotonin, the neurotransmitter in the brain that deals with both depression and drug craving. In addition to treating the physical and psychological aspects of craving, treatment should stress group counseling and peer pressure for compulsive amphetamine users, as these forms of therapy work well for this population. Sedatives such as flurazepam (Dalmane®), chloral hydrate, chlordiazepoxid (Librium®), phenobarbital, or even diazepam (Valium®) can be used, very carefully, on a short-term basis to treat anxiety or sleep disturbance problems. Antipsychotic medications such as haloperidol (Haldol®), chlorpromazine (Thorazine®), and others are also used to buffer the effects of the unbalanced dopamine level, the neurotransmitter that moderates paranoia and pleasurable sensation.

Non-medical Treatment in Methamphetamine Related Depression

Because of methamphetamine-induced hyperstimulation of pleasure pathways, this later leads to anhedonia. Former users have noted that they feel stupid or dull when they quit using methamphetamine. It is possible that daily administration of the amino acids L-tyrosine and L-tryptophane, precursors in the formation of neurotransmitters, can aid in the recovery process by making it easier for the body to reverse the depletion of dopamine, norepinephrine, and serotonin. Although studies involving the use of these amino acids have shown some success, this method of recovery has not been shown to be consistently effective. In addition, it has been shown that taking ascorbic acid prior to using methamphetamine may help reduce acute toxicity to the brain, as rats given the human equivalent of 5–10 g of ascorbic acid 30 min prior to methamphetamine dosage showed a reduced toxicity [26]. Yet this will likely be of little avail in solving the serious behavioral problems associated with methamphetamine use that create many of the problems the users experience.

Treatment Preventing Methamphetamine Relapse

To combat addiction, other forms of amphetamine such as dextroamphetamine are being used to break the addiction cycle, a method similar to methadone for heroin addicts. There are no publicly available drugs comparable to naloxone, which blocks opiate receptors and is therefore used in treating opiate dependence, for use with methamphetamine problems. However, experiments with some monoamine re-uptake inhibitors such as indatraline have been successful in blocking the action of methamphetamine [27]. Similar as in cocaine abuse, there are studies indicating that fluoxetine, bupropion and imipramine may reduce craving and improve adherence to treatment [24]. Research has also suggested that modafinil [28] and the potent opioid antagonist naltrexone [29] can help addicts quit methamphetamine use because it resulted in a significant reduction in craving and helped to stay abstinent.

Since phenteramine (2-methyl-1-phenylpropanin-2-amine) is a constitutional isomer of methamphetamine, it has been speculated that it may be effective in treating methamphetamine addiction. Although phenteramine is a central nervous stimulant that acts on dopamine and norepinephrine, it has not been reported to cause the same degree of euphoria that is associated with other amphetamines.

Amphetamine Derivatives as Appetite Suppressants

For decades, amphetamine stimulants have been used to ease hunger. Unfortunately, they are addictive and have adverse effects on blood pressure and heart health. Less addictive drugs related to amphetamines have not been proven to be very effective. Moreover, they produce serious side effects. Two of these, fenfluramine and phenteramine (called "fen-phen" when used in combination), caused damage to heart valves in too many users. Phenylpropanolamine, a decongestant, was available over the counter for appetite suppression. Although its effect was variable, a significant number of people developed a cerebral hemorrhage from its use. As a result it has been withdrawn from the market, even for treating nasal congestion. Also, reports are accumulating that ingestion of appetite suppressants (e.g., Dextro/Lexoamphetamine or Adderall®; Table 3) may results in cardiomyopathy [30]. In this report a patient is described who developed cardiomyopathy after receiving a therapeutic course of dextroamphetamine/amphetamine. The patient's cardiac function deteriorated to the point of heart failure, necessitating a heart transplantation. Cardiomyopathy associated with amphetamine use is a serious and potentially lethal condition. With early diagnosis, identification of the cause, and treatment, cardiomyopathy may be reversible. Because of such dangers of therapeutic use of amphetamines, as well as problems and assumptions, the U.S. Food and Drug Administration, removed several harmful appetite suppressants from the market.

Table 3 Some commonly used appetite suppressants, where metabolization results in the formation of either amphetamine (*right*) of methamphetamine (*left*)

Methamphetamine	Amphetamine
Amphetaminil	Benzphetamine
Clobenzorex	Deprenyl
Ethylamphetamine	Dimethylamphetamine
Fenetylline	Famprofazone
Fneproporex	Fencamine
Mefenorex	Furfenorex
Mesocarb	Selegiline
Prenylamine	
Selegiline	

References

1. Winslow BT, Voorhees KI, Pehl KA. Methamphetamine abuse. Am Fam Physician. 2007;76:1169–1174.
2. Nagai N. Kanyaku maou seibun kenkyuu seiseki (zoku). Yakugaku Zashi. 1893;13:801.
3. Müller-Bonn H. Pervitin, ein neues Analepticum. Med Welt. 1939;39:1315–1317.
4. Pellmont B. Vergleichende Untersuchungen über die Wirkungen von Coralin, Coffein und Pervitin auf psychsche und physische Leistung des ermüdeten und nicht-ermüdeten Menschen. Naunyn-Schmiedebergs Arch Exp Pathol Pharmacol. 1942;199:274–291.
5. Doyle D. Hitler's medical care. J Roy Coll Phys Edinburgh. 2005;35:75–82.
6. Kemper WR. Pervitin – Die Endsiegdroge ? Wach und leistungsstark durch Metamphetamin. In: Peier W, editor. Nazis on Speed – Drogen im 3 Reich. Löhrbach: Werner Pieper & The Grüne Kraft 1989. pp. 122–133.
7. Tamura M. Japan: stimulant epidemics past and present. In: Crime OoDa, editor. Geneva: United Nations Bulletin on Narcotics. 1989; 41(1 and 2):83–93.
8. Volkow ND, Chang L, Wang G-J, Fowler JS, Franceschi D, Sedler M, et al. Loss of dopamine transporters in methamphetamine abusers recovers with protracted abstinence. J Neurosci. 2001;21:9414–9418.
9. Bennett B, Hollingswort HC, Martin R, Harp J. Methamphetamine-induced alterations in dopamine transporter function. Brain Res. 1998;582:219–227.
10. Marx J. How stimulant drugs may calm hyperactivity. Science. 1999;283:306–309.
11. Garris PA, Kilpatrick M, Bunin MA, Michae lD, Walker QD, Wightman RM. Dissociation of dopamine release in the nucleus accumbens from intracranial self-stimulation [letter]. Nature. 1999;398:67–69.
12. McGregor C, Srisurapanont M, Jittiwutikarn J, Laobhripatr S, Wongtan T, White J. The nature, time course and severity of methamphetamine withdrawal. Addiction. 2005;100:1320–1329.
13. Rothman RB, et al. Amphetamine-type central nervous system release of dopamine and serotonin. Synapse. 2001;39:32–41.
14. Itzhak Y, Martin J, Ali S. Methamphetamine-induced dopaminergic neurotoxicity in mice: long-lasting sensitization to the locomotor stimulation and desensitization to the rewarding effects of methamphetamine. Prog Neuropsychopharm Biol Psychiat. 2002;26:1177–1183.
15. Davidson C, Gow AJ, Lee TH, Ellinwood EH. Methamphetamine neurotoxicity: necrotic and apoptotic mechanisms and relevance to human abuse and treatment. Brain Res Rev. 2001;36:1–22.
16. Bennett B, Hollingsworth C, Martin R, Harp J. Methamphetamine-induced alterations in dopamine transporter function. Brain Res. 1998;782:219–227.
17. Yamamoto B, Zhu W. The effects of methamphetamine on the production of free radicals and oxidative stress. J Pharmacol Exp Ther. 1998;287:107–114.

18. Yamamoto B, Zhu W. Relationship between temperature, dopaminergic neurotoxicity, and plasma drug concentrations in methamphetamine-treated squirrel monkeys. J Pharmacol Exp Ther. 2006;316:1210–1218.
19. Thomas DM, Walker PD, Benjamins JA, Geddes TJ, Kuhn DM. Methamphetamine neurotoxicity in dopamine nerve endings of the striatum is associated with microglial activation. J Pharmacol Exp Ther. 2004;311:1–7.
20. Reese EA, Bunzow JR, Arttamangkul S, Sonders MS, Grandy DK. Trace amine-associated receptor 1 displays species-dependent stereoselectivity for isomers of methamphetamine, amphetamine, and para-hydroxyamphetamine. J Pharmacol Exp Ther. 2007;321:178–186.
21. Lewin AH. Receptors of mammalian trace amines. AAPS J. 2006;8:E138–E45.
22. McGregor C, Srisurapanont M, Jittiwutikarn J, Laobhripatr S, Wongtan T, White J. The nature, time course and severity of methamphetamine withdrawal. Addiction. 2005;100:1320–1329.
23. Cook CE. Pyrolitic characteristics, pharmacokinetics, and bioavailability of smoked heroin, cocaine, phencyclidine, and methamphetamine. In: Miller MA, Kozel NJ, editors. Methamphetamine Abuse: Epidemiologic Issues and implications. DHHS Pub. No. (ADM) 91–1836. NIDA. Rockville, MD; 1991. pp. 6–23.
24. Winslow BT, Voorhees KI, Pehl KA. Methamphetamine abuse. Am Fam Physician. 2007;76:1169–1174.
25. Richards JR, Brofeldt BT. Patterns of tooth wear associated with methamphetamine use. J Peridontol. 2000;71:1371–1374.
26. Wagner GC, Carelli RM, Jarvis MF. Pretreatment with ascorbic acid attenuates the neurotoxic effects of methamphetamine in rats. Res Commun Chem Pathol. Pharmacol. 1985;47:221–228.
27. Rothman RB, et al. Neurochemical neutralization of methamphetamine with high-affinity nonselective inhibitors of biogenic amine transporters: a pharmacological strategy for treating stimulant abuse. Synapse. 2000;35:222–227.
28. Grabowski J, et al. Agonist-like, replacement pharmacotherapy for stimulant abuse and dependence. Addict Behav. 2004;29:1439–1464.
29. Jayaram-Lindström N, Hammarberg A, Beck O, Franck J. Naltrexone for the treatment of amphetamine dependence: a randomized, placebo-controlled trial. Am J Psychiatr. 2008;165:1442–1448.
30. Marks DH. Cardiomyopathy due to ingestion of adderall. Am J Ther. 2008;15:287–289.

Part III
MDMA (Ecstasy): from Psychotherapy to a Street Drug of Abuse

History of MDMA

The use and abuse of MDMA (3,4-methylendioxymeth-amphetamine), called *Ecstasy* or *XTC* on the East cost, or *Adam* on the West coast of the US, has significantly increased not only in California but also in other parts of the country and world wide. After being first synthesized for Merck by the German chemist Anton Köllisch at Darmstadt earlier in 1912 as a possible appetite suppressant, a patent was granted in 1914. Köllisch died in 1916 unaware of the impact his synthesis would have [1, 2]. However, MDMA remained an obscure amphetamine congener until the 1970, when psychedclic advocates used MDMA as a probe of consciousness [3], while a few psychotherapists used it as an adjunct to psychotherapy. MDMA began to be used therapeutically in the mid-1970's after the chemist Alexander Shulgin introduced it to psychotherapist Leo Zeff [4]. As Zeff and others advocated the use of MDMA, it developed a reputation for enhancing communication, reducing psychological defenses, and increasing capacity for introspection. However, no formal measures of these putative effects were made, and blinded or placebo-controlled trials were not conducted. A small number of therapists, including George Greer, Joseph Downing, and Philip Wolfson, used it in their practices until it was made illegal [5]. There even have been accounts that the military used MDMA for interrogating enemy spies.

The use of medications to gain access to repressed memories or to buffer emotional response to painful memories has a long tradition in psychiatry. When sedatives such as intravenous sodium thiopental (Pentothal®) are used, the procedure is sometimes called narcosynthesis. During the 1950's psychotherapists administered low doses of LSD [6] or MDA [7] to patients as an adjunct to psychotherapy. The term "psycholytic therapy" was used to describe psychotherapy augmented by psychedelic-type drugs. A psychedelic drug given in a high dose without psychotherapeutic assistance was generally termed *psychedelic therapy*. However, MDMA's therapeutic potential wasn't fully realised until 1976, when the American chemist Alexander Shulgin tried it on himself. He noted that its effect, "*an easily controlled altered state of consciousness with emotional and sensual overtones*", could be ideal for psychotherapy, as it induced a state of openness and trust without hallucination or paranoia. It quickly became known as a wonder drug, and began to be used widely in couple's therapy and for treating anxiety disorders. None of these tests was *empirical* in the scientific sense – no placebos, no follow-up testing – but anecdotally the results were almost entirely positive. Also, during the 1960's psychedelic therapy was under study for the treatment of alcoholism [8] and another option, which presently is under reinvestigation, was for post-traumatic stress syndrome (PTSD).

Psychotherapists have administered MDMA as an adjunct to therapy in much the same way as low-dose LSD in the 1980's. The history of MDMA's use as a psychotherapeutic adjunct parallels that of LSD and involves many of the same advocates. But before there was a general consensus about its efficacy, other researchers began giving it to patients in a non-research setting, with some therapists administering it responsibly. Others administering the agent to their patients under conditions, where the therapeutic intent was questionable. Already in the 1980's Greer reported improvement in psychiatric disorders and mood in his patients. Some patients reported positive changes in attitude, increased self-esteem, more acceptance of negative emotional experiences, and belief of changes, which persisted from weeks to years following the session [5]. The efficacy of MDMA-augmented psychotherapy was not established. As a consequence of being placed in Schedule I of the US Controlled Substance Act in 1985, MDMA psychotherapy research was stopped; however, illicit recreational use was unabated.

In 1980's the street drug culture, especially the *ravers* discovered MDMA as a drug to increase the feeling of empathy, the *touching within* and the enhancement of senses. A rave (or rave party), is a term, in use since that time, to describe dance parties (often all-night events) where DJs and other performers play electronic dance music, sometimes referred to as *rave music*, with the accompaniment of laser light shows, projected images, and artificial fog. Rave parties often, not always, are associated with the use of so called *club drugs* such as ecstasy, cocaine, amphetamines and, more recently, ketamine and the designer drug 2C-B (4-Bromo-2,5-dimethoxy-phenethylamine; Fig. 1), a synthetic agent originally synthesized by A. Shulgin in 1975 [9].

Similar as in the US, rave-like dances were held in the early 1980s in the Ecstasy-fueled club scene parties in England. However, it was not until the mid to late 1980s that a wave of psychedelic and other electronic dance music, most notably acid house and techno, emerged and caught on in these clubs, warehouses and free-parties around the larger cities. These early raves were called the *Acid House Summers*. They were mainstream events that attracted thousands of people (up to 25,000) instead of the 4,000 that came to earlier parties (Fig. 2). In 1988–1989, raves were similar to football matches in that they provided a setting for working-class unification.

Because of the growing misuse of MDMA at these rave parties the DEA evoked emergency scheduling provisions of the Controlled Substances Act on May 31, 1985 and placed MDMA temporarily in Schedule I (=no medical use, but high abuse potential). UK laws were even tighter: MDMA is illegal under the 1971 Misuse of Drugs Act, it was categorized as class A in 1977, carrying a sentence of up to 7 years for possession.

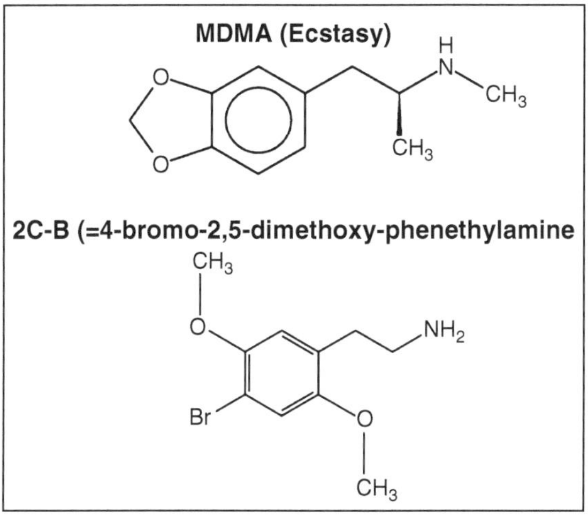

Fig. 1 Molecular structure of MDMA (top) Ecstasy, Adam, or X and 2C-B (bottom) commonly known as "Nexus", "Venus", "Bromo", or "Eros"

Fig. 2 Ecstasy (XTC) or MDMA primarily a drug of the young raver generation, is now a cult drug of the young people (<34 years of age) who for the sake of an extended feeling for empathy, harmony, and extreme happiness regularly abuse the drug at weekend parties and in discotheques

Incidence of Illicit Use of Ecstasy

Clandestine laboratories operating throughout Western Europe, primarily the Netherlands and Belgium, manufacture significant quantities of the drug in tablet, capsule, or powder form. Although the vast majority of ecstasy or 3,4-methylene-dioxymethamphetamine (MDMA) consumed domestically is produced in Europe, a limited number of MDMA labs operate in the United States. In addition, in recent years, Israeli organized crime syndicates, some composed of Russian exiles associated with Russian organized crime syndicates, have forged relationships with Western European traffickers and gained control over a significant share of the European market. The Israeli syndicates are currently the primary source to US distribution groups.

Overseas MDMA trafficking organizations smuggle the drug in shipments of 10,000 or more tablets via express mail services, couriers aboard commercial airline flights, or, more recently, through air freight shipments from several major European cities to cities in the United States (Fig. 3). The drug is sold in bulk quantity at the mid-wholesale level in the United States for approximately $8 per dosage unit. The retail price of MDMA sold in clubs in the United States remains steady at $20–$30 per dosage unit. MDMA traffickers consistently use brand names and logos as marketing tools to distinguish their product from that of competitors. The logos are produced to coincide with holidays or special events. Among the more popular logos are butterflies, lightning bolts, and four-leaf clovers (Fig. 4).

Medical Use of Ecstasy in Assisting Psychotherapy

Aside from illicit use in discotheques, the MDMA used in a clinical study, is the pure chemical compound, not the black-market ecstasy bought by recreational users. This is an important point to be considered, because a lot of ecstasy pills are not MDMA at all. They may be amphetamines, or unknown pharmaceuticals, or they can be diluted with almost any drug in pill or powder form. That's when the user magnifies risks associated with taking a drug that's already toxic. In addition, people use it irresponsibly, mixing it with other drugs (e.g. alcohol), not drinking enough water or drinking too much. Presently a phase II clinical trial study is under

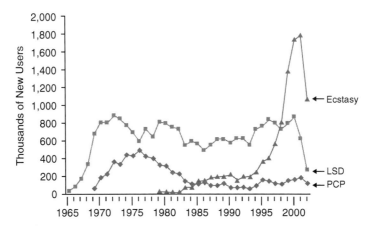

Fig. 3 Annual numbers of new users of ecstasy, LSD, and PCP from 1965 to 2002

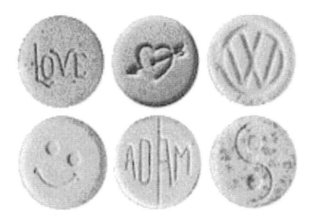

Fig. 4 Typical examples of MDMA tablets with different logos used for oral consumption as they are sold at discotheques

way to evaluate the safety and efficacy of MDMA-assisted psychotherapy in subjects with chronic posttraumatic stress disorder. Posttraumatic stress disorder (PTSD) occurs in response to a traumatic event, and is most likely to occur following an event involving perceived personal threat, such as rape or physical assault. PTSD presents a public health problem that causes a great deal of suffering and accounts for a significant portion of health care costs. The study is directed to examine whether two sessions of MDMA-assisted psychotherapy can be safely administered to participants with PTSD. It is also evaluated, if MDMA-assisted psychotherapy, when compared with placebo-assisted therapy, will reduce PTSD symptoms after

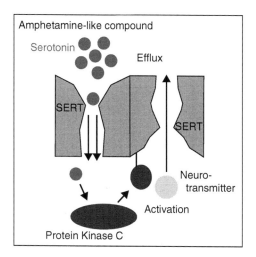

Fig. 5 Postulated effect of the amphetamine-like compound MDMA on the serotonin transporter SERT system followed by a reuptake of the transmitter serotonin and an increase of translocation of protein kinase C (PKC)-dependent phosphorylation from the cytosol to the membrane resulting in an impulse-dependent dopamine release activation (Adapted from [13])

each session and 2 months after the second session. The study is sponsored by the non-profit Multidisciplinary Association for Psychedelic Studies (MAPS), and has been approved in the United States, Switzerland and Israel.

Besides such a potential use in psychotherapy, MDMA also is being evaluated for a potential benefit in palliative care. In addition, MDMA has been shown to have favorable effects in patients with severe Parkinsonism [10]. Several lines of evidence support the notion of MDMA being a unique and new option as an antiparkinson drug [11]. As to its mode of action in Parkinsonism it is being discussed that MDMA acts via the serotonin transporter (SERT) and protein kinase C (PKC)-mediated signaling pathways, thus modulating MDMA-induced dopamine release from mesocorticolimbic and nigrostriatal neurons [12]. Being an integral membrane protein, SERT transports the neurotransmitter serotonin from synaptic spaces into presynaptic neurons (Fig. 5). This transport of serotonin by the SERT protein terminates the action of serotonin and recycles it in a sodium-dependent manner.

Pharmacology of Ecstasy (MDMA)

Like mescaline, the primary psychoactive ingredient in peyote, MDMA belongs to a group of ring-substituted amphetamine congeners, commonly called "methoxylated amphetamines" (Fig. 6).

The drug MDMA is a potent indirect monoaminergic agonist, which is thought to act by both increasing the release and inhibiting the reuptake of serotonin and, to a lesser extent, dopamine [14]. Serotonin is involved in the regulation of a variety of behavioral functions, including mood, anxiety, aggression, appetite, and sleep. Dopamine is the primary neuro-transmitter of the "reward pathway" and is involved in motivational processes such as reward and reinforcement. Norepinephrine has important roles in the process of attention and arousal. In vitro studies have shown, that MDMA causes release of serotonin, dopamine, and norepinephrine from synaptosomes [15, 16] and from rat brain slices [14]. In freely moving rats, MDMA increases both serotonin and dopamine release in the caudate [14]. In a similar study, MDMA increased dopamine release in awake rats, resulting in region-, time-, and dose-dependent behavior [17]. In rat brain synaptosomes, MDMA inhibited the uptake of serotonin and norepinephrine and, to a lesser extent, dopamine. The local administration of MDMA to the rat nucleus accumbens resulted in increases in the extracellular levels of both serotonin and dopamine in this region, which is part of the reward pathway activated by other abused substances such as amphetamine and cocaine. These actions in the nucleus accumbens may account for the euphoric effects produced by MDMA.

In addition to causing the release of serotonin and inhibiting its reuptake, MDMA may have direct agonist effects on serotonin and dopamine receptors [18]. It has affinities for a broad range of neurotransmitter recognition sites and may act at both serotonin subreceptors, 5-HT2_A and 5-HT2_C [19]. Selective serotonin reuptake inhibitors (SSRI) such as fluoxetine and citalopram block the release of serotonin induced by MDMA, both in vitro and in vivo [20]. They also reportedly block the subjective effects produced by MDMA in humans. Consequently, the release of serotonin by MDMA may be dependent on the serotonin transporter SERT. MDMA shows different potencies for the neurotransmitter systems than either amphetamines or hallucinogens (Fig. 7).

Mescaline (=3,4,5-trimethoxyphenethylamine)

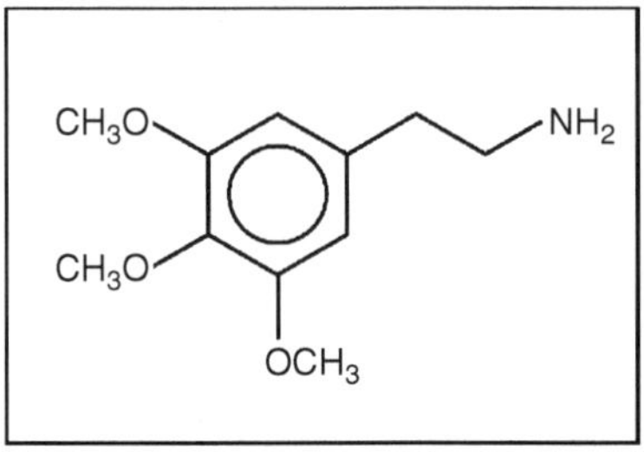

MDA (=3,4-methylene-dioxyamphetamine)

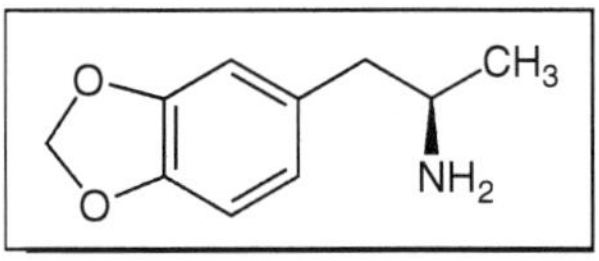

MDEA (or MDE=methelene-dioxy-ethylamphetamine

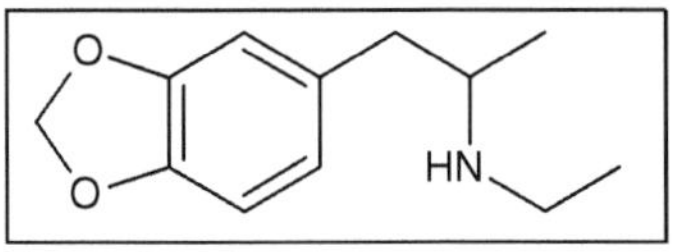

MDMA (=3,4-methylene-dioxymethamphetamine)

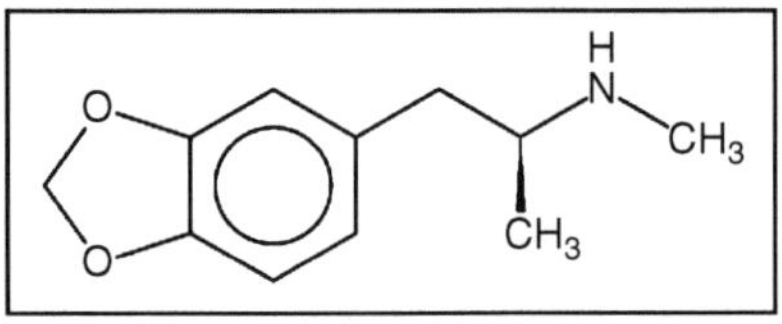

Fig. 6 Similarity in the molecular structure of the drug ecstasy (MDMA), MDA, MDEA and their relation to mescaline, a hallucinogenic found in the cacti Peyote (lophophora williamsii), San Pedro (trichocerus pachanoi) and the Peruvian Torch (trichocerus peruvianus). MDEA is sometimes sold as a substitute for ecstasy on the black market and MDA is a metabolite of MDMA

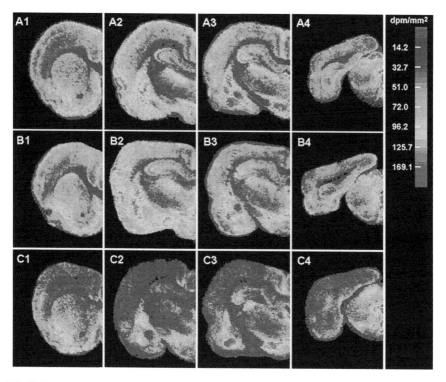

Fig. 7 Receptor autoradiography using the radioisotope [125I]RTI-55 to label serotonin transporters in control Wistar rats (**a**), rats treated with a low dose of MDMA (**b**) and rats treated with a high dose of MDMA (**c**). Note the significant decrease in serotonin transporters (SERT) observed in both high and low dose MDMA groups in cortical areas and the medial hypothalamus, while decreases in the lateral septum, amygdaloid nuclei, hippocampus and thalamic nuclei are seen only in the high dose group [19]

Pharmacological Effects of MDMA in Man

Onset of Action and Duration of Action

Ecstasy or MDMA in its pure form is a white crystalline powder. It is usually seen in capsule form, in pressed pills (Fig. 4), or as loose powder. Common routes of administration are swallowing or snorting, although it can be smoked or injected as well. MDMA in the US and Europe is on Schedule I of controlled substances, and it is illegal to manufacture, possess, or sell it in the United States or other countries, who have similar laws.

Usual doses of MDMA in humans range from around 80–60 mg orally, while lower doses (40–60 mg) are used to assist meditation, and in psychotherapy. Subjective effects peak between 90 min and 2 h after ingestion of MDMA and return to baseline approximately 4 h after ingestion. A benchmark standard dose is often considered to be 2 mg of MDMA per kg of body weight, though response to the drug is not strictly proportional to body weight. When MDMA is taken orally, the effects manifest about 30–45 min later; snorting, smoking or injecting produces a much quicker onset (Table 1). The primary effects usually reach a plateau 1 h after taking the dose. It stays there for some 2 h, then start tapering gradually being over by four to 6 h. The psychological effects largely dissipate in 3 h with the exception of some residual sympathomimetic effects, which may last up to 5 h. Secondary effects (afterglow) may be felt for days, and tertiary psychological effects (e.g. improved outlook) may last indefinitely [3].

The Psychological Effects of MDMA

The psychological effects are difficult to describe, since they vary individually. The major ones are:

Entactogenesis ("Touching Within")

Owing to its individual biochemical profile and the subjective effects it produces in humans, MDMA has been called an entactogen, which means, "producing a touching within". This is a generalized feeling that all is right and good with the world. People on MDMA often describe feeling "at peace" or experiencing a generalized "happy" feeling. Also, common everyday things may seem to be abnormally beautiful or interesting. Alexander Shulgin reported that mountains that he had observed many times before appeared to be so beautiful that he could barely stand looking at them [4].

Empathogenesis

Empathogenesis is a feeling of emotional closeness to others and to one's self, coupled with a breakdown of personal communication barriers. People on MDMA report feeling much more at ease talking to others and that any inhibitions that one may have with regard to *"opening up"* to others may be reduced or even eliminated. This effect is partially responsible for MDMA being known as a *"hug drug"*, where the increased emotional closeness makes personal contact rewarding. Many people use MDMA primarily for this effect, reporting that it makes potentially awkward or uncomfortable social situations much more easily to deal with. Conversation is facilitated. It seems like you know exactly what to say and when to say it, the filter between had been removed what you want to express, not knowing it even existed.

Enhancement of Senses

MDMA can significantly enhance (sometimes distort) the senses – touch, proprioception, vision, taste, and smell. MDMA users can sometimes be seen running their hands over differently textured objects repeatedly, tasting and smelling various foods/drinks. This effect also contributes to the "hug drug" effect because of the novel feeling of running one's hands over skin and having one's skin rubbed by someone else's hands.

Before it was made illegal, it was because of this *"opening up"* that MDMA was gaining a reputation among the psychiatric community as a valuable therapeutic tool. People under its influence often report seeing their lives in a whole new light. Some of the problems the person didn't even know he had, all of a sudden became obvious as they seemed to be the source, the nature and sometimes even the solution of all personal difficulties. Surfacing of repressed memories also has been reported. Despite the legal risks surrounding Schedule I drugs, some therapists are still using MDMA in their practices [21].

Most people find the MDMA state so valuable by itself that it is not clear if there is much to be gained from combining MDMA with other substances, though the combination of MDMA with LSD seems to have a strong following. Further, combining drugs ("polydrug use" and "polydrug abuse") complicates the medical and behavioral safety picture. For this reason, it is not a recommended practice.

Abuse Liability and Psychomotor Performance

Although it appeared that MDMA has a low abuse potential and no significant toxicity was observed when the drug was use as an adjunct in psychotherapy, as might be expected, significant MDMA toxicity does occur when chronic high doses are used. This, for instance, is seen in patients with a history of drug abuse who seem to be particularly susceptible to MDMA abuse. For example, a patient who was on cocaine stopped his abuse, but began using 250–750 mg of MDMA daily instead.

The schedule I action by the DEA, was to avoid an imminent hazard to the public safety, i.e. the possibility that MDMA might cause irreversible brain damage. Although no specific animal or human research on MDMA supported this notion of brain damage, the potential was based entirely on the comparison of MDMA with the structural congener MDA (=3,4-methylene-dioxyamphetamine). Researchers had noticed a degradation of nerve terminals in rats following 4 days of subcutaneous administration of MDA in a dosage of 5 mg/day or greater given every 12 h [22].

In controlled studies, MDMA produced marked feelings of euphoria and well-being and possessed amphetamine-like properties [23]. Assessments of positive psychologic states have been shown to increase with increasing dosage. MDMA produces mild changes in perceptions but does not commonly cause hallucinations or psychotic episodes [23]. It also produces moderate derealization and depersonalization, as well as anxiety without marked increases in psychomotor drive. Since MDMA appears to involve serotonin stimulation, it is suggested that chronic abuse can lead to long-term damage of serotonin-containing neurons, causing persistent memory impairments. Other functional aspects of serotonin – mood, impulse control, sleep cycles – may be affected, and it is now believed that serotonergic dysfunction may persist for many years or may even become permanent [24].

The Physical Effects of MDMA

The physical effects of usual doses of MDMA are subtle and variable; some users report dryness of mouth, jaw clenching, teeth grinding, nystagmus, sweating, or nausea (Table 1). During the peak effect there is a moderate increase in heart rate and blood pressure. Others report feelings of profound physical relaxation. At higher doses (overdoses), the physical effects of MDMA resemble those of amphetamines characterized by a fast or pounding heartbeat, sweating, dizziness, restlessness, etc.

Table 1 Summary of the acute effects of MDMA

After the ingestion of 5–150 mg of MDMA the following effects can be observed:

1. Perception of increased trust and intimacy with others
2. Enhanced communication
3. Increased insight into personal pattern of personal behavior
4. Self-awareness
5. Feeling of warmth, freshness and love
6. Euphoria
7. Sensations of being more self-aware, alive, and at peace

Common negative effects, however, can also be noticed which are characterized by:

1. Anxiety
2. Teeth grinding
3. Cheek-biting
4. Anorexia
5. Subsensitive sensation of being cold
6. Insomia
7. Reduced feeling of thirst

Cardiovascular Effects of MDMA

Ingestion of MDMA 75 mg and 125 mg causes an increase in blood pressure and heart rate, which occurs maximally at 90 min and 60 min, respectively. Marked increases in blood pressure and heart rate were also seen after doses of MDMA 0.25–1.0 and 1.7 mg/kg, respectively [20]. In the latter investigation, peak increases in blood pressure occurred 2 h after drug administration, and 12 of the 13 subjects had a peak blood pressure of 160/100 mmHg, whereas blood pressure in the last patient peaked at 240/145 mmHg [25].

Neuroendocrine Effects of MDMA

Plasma cortisol levels are significantly increased after ingesting MDMA at both 75 mg and 125 mg, and the 125 mg dose causes significant elevations in prolactin levels as well. The increases in cortisol and prolactin levels reach a peak at 2 h after MDMA administration [15]. Other findings state that adrenocorticotropic hormone (ACTH) and prolactin concentrations are increased after the oral administration of MDMA 0.75–1.0 mg/kg and that increases in serotonergic function increase both cortisol and prolactin levels. For example, fenfluramine, a drug that causes serotonin release, produced dose-related increases in the concentrations of both cortisol and prolactin [26]. In addition to cortisol and prolactin, plasma vasopressin (arginine vasopressin = AVP) was significantly elevated 1–2 h after MDMA administration, which may be accompanied by a small decrease in serum sodium levels and unchanged cortisol levels. Therefore, the slight hyponatremic effect could be related to the ability of MDMA to release AVP rather than a stress response, as elevated AVP levels were accompanied by unchanged cortisol levels. However, similar doses caused significant elevations in ACTH in a study by a different group of researchers [27].

Ocular Effects of MDMA

Ingestion of MDMA 75 mg and 125 mg produces mydriasis, with a maximal change in pupillary diameter occurring 1–2 h after drug administration. Furthermore, MDMA 125 mg produces significant esophoria with a tendency for the eyes to turn inward.

Lethal Effects of MDMA

The lethality of a compound (LD_{50}) is defined as the dose of a drug that will kill 50% of the animals receiving that dose. Work done by the US Army in 1953–1954 compared the 24 h LD_{50} among five different animals (mice, rats, guinea pigs, dogs, and monkeys). The LD_{50} was 49 mg/kg in rats, 14 mg/kg in dogs, and 22 mg/kg in rhesus monkeys [28]. These findings indicate that MDMA causes a significant dose-dependent toxic reaction and death in many animals. It is, however, difficult to extrapolate these data to humans because the animal data were obtained after intravenous or intraperitoneal administration. Since MDMA is taken orally, its oral bioavailability in humans is unknown.

Effects in Chronic Use of MDMA

In order to have a predictable response to MDMA, one needs to have good quality MDMA. The only way to maximize the chance of getting the pure agent is to know and trust the supplier. This is because many times the synthesis of MDMA is done in a home laboratory with little expertise, resulting in an accumulation of so called by-products during the chemical reactions. Note that MDMA is not known for causing strong visual distortions. If a person takes some "*MDMA*" and notices that a predominant effect is visual distortion, then it probably was not pure MDMA, or MDMA at all. Most users of MDMA who have taken the drug many times report that after some number of sessions, varying by person from a few to a dozen, the desirable effects of the drug are no longer as pronounced. Persons who had been using MDMA a dozen times separated by weeks to months, noted the drop-off, waited 3 years, and then again tried an ordinary dose of high-quality MDMA. They found that the annoyance of the physical side effects outweighed the greatly diminished positive effects. Others who have had fifty or more MDMA sessions still find them to be worthwhile.

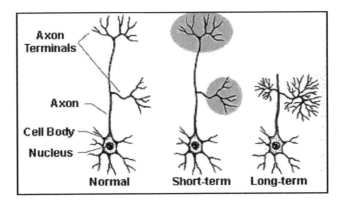

Fig. 8 The alleged effect of MDMA on serotonin neurons in the monkey brain, which later turned out to be methamphetamine (Adapted from [30])

Chronic MDMA Use and Neurotoxicity

This MDMA drop-off effect might be explained by a psychological mechanism: loss of novelty. On the other hand, people who have experienced a drop-off effect after MDMA generally report that there is not a similar drop-off in the effects of other psychedelics with which they are equally or more experienced, e.g. LSD and Dimethyltryptamin (DMT). It is now discussed, that the drop-off is caused by lasting neurophysiological or neurological *changes* to the brain from exposure to MDMA. This has been emphasized by PET studies in animals (Fig. 8) which, however, later turned out to be due to an erroneous use of the close relative methamphetamine and not MDMA.

As noted, a primary psychological effect of MDMA is to make the user feel *safe*, at peace with the world, pleasantly reconciled to things as they are. This can remarkably diminish one's ability to make sound judgements while on the drug. Another danger stems from MDMA's lessening of the awareness of pain. This is either through chemical, or through psychological analgesia. Combined with the extra energy the drug gives, it becomes easy to sustain bruises, blisters, or other bodily damage from extensive dancing, hiking, climbing, etc., without noticing it until after the damage is done. Under MDMA, it may seem *correct* to make immediate changes in relationships (increasing or decreasing commitment) of all kinds. The fresh points of view appreciated during an MDMA experience are one of the drug's most prized benefits. It is probably unwise to actually make lasting relationship changes until the person has a chance to see how he feels about them after the drug and its afterglow wear off.

Another repetitively claimed effects of MDMA use are the lowered brain serotonin levels in the brain. One study [29] found no evidence for this, but at least two others [30, 31] have found significantly reduced serotonin metabolite levels, the more recent study showing a 30% average difference between the control group of non-MDMA users and the experimental group consisting of people who had used MDMA about

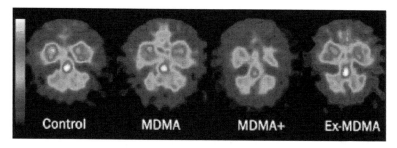

Fig. 9 SPECT-analysis of MDMA moderate and excessive use depicting the density of serotonin and its recovery after termination (Adapted from [35])

75 times each, on average. One however should note though, that some of these studies used psychiatric patients or *"polydrug abusers"* which are not representative user samples.

The study in animals also showed that the monkeys who were given MDMA chronically had damage to the serotonin-containing nerve cells. This damage was still visible 7 years later [30]. Areas that were especially affected were the frontal lobe of the cerebral cortex, an area in the front part of the brain that is used in thinking, and the hippocampus, an area deep in the brain that involves memory. Although damage was still observed 7 years later, it was less severe than when it was observed 2 weeks after drug use. This suggests that some neuron regeneration could have occurred (Fig. 9), as demonstrated by SPECT analysis in former MDMA users [32].

While anecdotal evidence from years of use suggests that this is not of much concern for most people, some users, however, report periods of depression after using MDMA, on rare occasion severe depression. Considering that a primary action of many antidepressant drugs (MAOIs, SSRIs) is to increase brain serotonin levels, a connection between MDMA use and subsequent depression is reasonable. Psychological factors, such as sadness at returning to an ordinary state of consciousness after ecstasy, may also account for feeling depressed for a while. In any event, most users report the opposite: feelings of well-being or gentle euphoria in the days following a MDMA session.

To get a better understanding of why the serotonin system may be critical for some individuals and less so for others, there is solid experimental evidence that MDMA, administered in large doses and/or repeatedly, causes partial loss of serotonergic neurons in laboratory animals. Uncertain is whether this loss is permanent, reversible, or not important, especially as one study found a nearly 100% recovery within a year. In another study [31], non-human primates were dosed with MDMA and their brains were examined for morphological changes. Ricaurte found that there was no effect after 2.5 mg/kg oral doses given every 2 weeks, for a total of eight doses. But after a single oral dose of 5 mg/kg, he observed a 20% reduction in serotonin and its metabolite 5-hydroxyindoleacetic acid (5-HIAA), only in the thalamus and hypothalamus. There appeared to be some regeneration over time, not necessarily complete, and also some "collateral sprouting" - growth of other types of neurons in the reduced serotonin areas (Fig. 8).

In all of the animal studies, even when there are quite large serotonin system reductions (up to 90% in high MDMA dose rat studies), no behavioral deficits are observed. It is also uncertain how these studies would extrapolate to humans. The human brain may well be more or less sensitive, or sensitive in different areas, compared with other animals. In any case, what is known is that there are no reported cases that link behavior changes in humans with MDMA-induced serotonin system changes or neuronal loss. And, the long-term human behavior changes that are noted (in studies and from anecdotal case reports) are generally regarded as positive, such as lowered impulsiveness and hostility, improved social/interpersonal relations, changes in religious/spiritual orientation or practice, etc. One of the reasons so little is known about the lasting effects of MDMA on the human brain is, that few subjects have recorded their drug use history, and then volunteered their brains for post-mortem study.

Findings of a putative MDMA-related neuronal degeneration were underlined in the fall of 2002 by George Ricaurte, the *"Dark Prince of suspect science"*, who published an article in the prestigious journal Science that claimed that a *"common recreational dose"* of MDMA caused severe damage to the dopamine systems of monkeys [33]. Since Parkinson's disease involves the death of dopamine neurons, extensive damage to the dopamine axons could presumably cause symptoms similar to Parkinson's (Fig. 8). Following this paper, new anti-MDMA laws were passed, including the infamous RAVE Act. This made it illegal to have a party if you *knew* that some people would be using drugs such as MDMA. As a consequence, all studies which had been on the verge of finally having its MDMA post-traumatic stress disorder research approved, was stopped in the face of the apparent new evidence of MDMA's dangers. However, more than a year later, facing increasing questions from critics, the postulated *"evidence"* collapsed when Ricaurte *sheepishly* admitted that the experiment never really happened. The monkeys had actually been given massive doses of methamphetamine, and not MDMA !!! He also admitted that no matter what they had tried, they had been unable to damage monkey's dopamine systems with real MDMA. As a result of this his paper was retracted and NIDA (=National Institute of Drug Abuse, the sponsor of the studies) was asked under the Freedom of Information Act (FOIA) to disclose these studies and the previous ones, which also suggests that the potential risks of MDMA neurotoxicity have been exaggerated by government officials and government-funded scientists in a manner that has been used to support prohibitionist policies [34].

Contrary to Ricaurte's findings, research conducted at Duke University Medical Center has shown that MDMA is the most effective of 60 drugs tested in reversing the symptoms of Parkinson's disease [36]. Such concern over the research of Ricaurte reawakened the ecstasy neurotoxicity debate [37], and one may speculate, why previous studies of his group constantly misrepresented findings, including a deliberate failure to publish studies that found no effect of single low doses of oral MDMA on serotonin neurons.

Another study using the radioligand tracer (+)-McN5652 demonstrated different binding in subjects with a dissimilar history of drug use (i.e. current ecstasy use, previous ecstasy use, polydrug non-MDMA use or no drug use). Activity for (+)-McN5652 binding in the mesencephalon were lowest in current ecstasy users compared to all other groups with the following gradation: current ecstasy users < former ecstasy users = polydrug users = non-drug user controls. Thalamus (+)-McN5652 activity were lower in current ecstasy users than in either polydrug users or non-drug user controls demonstrating a gradation of current ecstasy users < polydrug users = non-drug users. Caudate (+)-McN5652 values were lower in current ecstasy users than in polydrug user controls (current ecstasy users < polydrug users) a difference that never reached statistical significance. If ligand binding distribution volume ratios serve as accurate measures of serotonin transporter site density, then the study findings indicate that continued regular, heavy ecstasy use lowered serotonin transporter availability, but that formerly regular, heavy ecstasy users no longer had lower serotonin transporter availability [38]. These findings suggest that repeated, heavy MDMA use (along with regular use of other substances) is associated with a temporary reduction in brain serotonin transporter availability. Changes in serotonin transporter availability may arise from down-regulation of serotonin function, harm to serotonin axons or both causes, but effects are not permanent. However, effects on cortical areas cannot be measured with (+)-McN5652, and at least one study in non-human primates has found reduced brain serotonin 7 years after a repeated-dose regimen [39].

Also, some users of MDMA report an apparent decrease in resistance to disease, especially with frequent MDMA use. It is unknown how much of this may be due to the pharmacological effect of MDMA, i.e. staying up all night and dancing, lack of sleep, to increased physical contact with people with colds, to suppressed appetite and poor nutrition, etc.

It also has been hypothesized that some of the neurotoxic actions of MDMA may result from quinones formed from the metabolism of dihydrometamphetamine (DHMA), which can combine with glutathione and other thiol compounds [40, 41].

Table 2 Suggested antioxydative nutrients for blocking phenethylamine damage (Adapted from [47])

Nutrient	Prevention dose	Therapeutic dose	Form
β-Carotene	5 mg	15 mg	Supplements of other carotinoids (e.g. glycopene)
Bioflavonoids	2 g	6 g	Mixed bioflavonoids
Coenzyme Q10	100 mg	300 mg	Tablet, Spray
Ascorbic acid	2–4 g	6–12 g	L-free acid/calcium, magnesium salt
L-Carnitine	1 g	3 g	L-Carnitine HCl or L-Carnitine Mg-citrate
N-Acetylcysteine	2 g	6 g	Only as L-Cysteine
Selenium	250 μg	500 μg	As Selenite or Selenomethionine
Vitamine E	1000 IU	3000 IU	Form not critical

A 6-hydroxydopamine analog is formed by the aromatic hydroxylation and demethylenation of MDMA that also could be neurotoxic [41, 42]. Catecholamines formed from MDMA, such as DHMA, are highly polar compounds that cannot cross the blood-brain barrier. However, these highly polar compounds have been detected in the brain after peripheral administration of MDMA, indicating that some MDMA metabolism may occur in the brain [43].

Concern about the possibility of serotonin level or serotonin system changes in humans with therapeutic doses of MDMA, lead some researchers to surmise, that changes can be lessened or prevented by taking antioxidants [44–46] (Table 2).

In addition, fluoxetine (Prozac®), when taken 0–6 h after MDMA, has been proclaimed as a preventive measure for serotonin system reduction [48]. One might speculate, that other SSRI drugs (Zoloft®, Paxil®) may be effective too. However, some people report that Prozac® taken before or in the early part of an MDMA experience lessens some of the desirable effects of the ecstasy. When taken together, the information is that there are a great many questions unanswered by research. Thus, a conservative, prudent assumption is that the risk of some kind of subtle neurological "*damage*" in humans from MDMA use is not zero. However, given any non-zero risk, it makes sense to examine the benefit side of the equation. Before not having more data available on the possible merits of MDMA, the agent should only be used if some tangible positive outcome may be expected and if the risk is worthwhile.

Pharmacokinetics of MDMA

Effects of MDMA usually begin 30–45 min after oral administration of a 75–150 mg dose with peak effects occurring 60–90 min after ingestion, which start to diminish after 2 h, however lasting up to 8 h (Table 3). The time to maximum concentration (Tmax) is 2 h after oral ingestion of MDMA 50, 75, or 125 mg. The half-life shows little variation after a wide range of doses. After a 50-, 75-, or 125-mg dose, the half-life is 8 h [49]. Other studies found the half-life to be 9.53 h after a 75-mg dose and 9.12 h after a 125-mg dose [50]. The maximum concentration (Cmax) after oral ingestion appears to be dose dependent. A Cmax of 105.6 ng/ml was reported in a single subject who took a 50-mg dose [51], whereas a Cmax of 330 ng/ml was found in another subject who took MDMA 135 mg [52]. In a group of eight subjects, the Cmax values after ingestion of MDMA 75 mg and 125 mg were 126.5 and 226.3 ng/ml, respectively [50], whereas in another group of eight subjects, Cmax values of 130.9 and 236.4 ng/ml were obtained after ingestion of MDMA 75 mg and 125 mg, respectively [49]. These studies indicate, that the Cmax exhibits a slightly greater-than-expected increase compared with the increase in dose. According to these observations, and following a usual recreational dose of 100–150 mg, the Cmax should be in the range of 200–300 ng/ml. The area under the concentration-time curve (AUC) data from these studies also suggests nonlinearity. The AUC measured over 24 h after ingestion of a 125-mg dose (2,235.9 µg/l/h) is more than twice the AUC after ingestion of a 75-mg dose (995.4 µg/l/h [50]. Nonlinearity is further supported by other evidence, in which the dose ratio of MDMA was 1:3 (50 mg and 150 mg), whereas the AUC ratio over 24 h after ingestion was greater than 1:10. The authors suggested that the non-renal clearance of MDMA is dose dependent.

Primary Metabolism of MDMA

The primary metabolic pathways for MDMA have been elucidated, with a number of metabolites identified in both animals and humans (Fig. 10). The main metabolic pathway appears to be demethylation to the catechol metabolite 3,4-dihydroxy-methamphetamine (DHMA; also called N-methyl-α-methyldopamine) [40].

Table 3 Summary of the physicochemical properties of MDMA

Name	MDMA, Ecstasy, E, XTC, Adam
Appearance	White, crystalline powder
Chemical names	3,4-methylenedioxymethyl-amphetamine
Alternate chemical names	3,4-methylenedioxy-N-methyl-amphetamine
	N,α-dimethyl-1,3-benzodioxole-5-ethanamine,
	Methylenedioxy-N-methyl-amphetamine,
	N-methyl-3,4-methylenedioxy-phenyl-isopropylamine
Chemical formula	$C_{11}H_{15}NO_2$
Molecular weight	193.25
Melting point	147–148°C (Hydrochloride crystals from isopropanol/n-hexane)
Melting point	152–153°C (Hydrochloride crystals from isopropanol/ether)
Solubility	Good solubility in water as a salt
LD_{50}	(mice) 97 mg/kg i.p.
LD_{50}	(rats) 49 mg/kg i.p.
LD_{50}	(guinea pigs) 98 mg/kg i.p.
Metabolism	Liver, N-demethylation and O-dealcylation, conjugation with glucoronide, 7% formation to MDA
Mean distribution time = onset of action	15 minutes
Mean plasma half-life = duration of action	4–6 hours

DHMA is the major metabolite of MDMA in rat liver [40, 53] and in rat brain microsomes [54]. Hydroxy-methoxy-methamphetamine (HMMA), one of the many metabolites of MDMA metabolism (Fig. 10), was the major product in plasma at lower doses, whereas MDMA was the predominant product at higher dose. This resulted in a disproportionate increase in plasma AUC and an increase in the proportion of MDMA excreted in the urine as the dose increased [55]. It is possible that demethylation may be inhibited as MDMA accumulates or one of the MDMA metabolites may inhibit cytochrome P450 (CYP) 2D6, which is responsible for a substantial proportion of MDMA nonrenal clearance. Alternatively, there might be an increase in the fraction of drug bioavailable as the dose increases [55]. Unfortunately, to our knowledge, the oral bioavailability of MDMA has not been determined in humans.

Microsomes from yeast expressing human CYP2D6 [41, 56] as well as human liver microsomes [57] demethylate MDMA also to the metabolite DHMA. If CYP2D6 is the isoenzyme responsible for the majority of MDMA metabolism in humans, then poor metabolizers could be sensitive to the acute physiologic effects of MDMA, but less prone to any long-term toxic effects of MDMA arising from metabolites. However, case reports have indicated that fatal MDMA intoxications have occurred in patients who were shown to be CYP2D6 extensive metabolizers [58], and it also has been shown in vivo that in the absence of functional CYP2D6 a considerable amount of metabolism of MDMA analogs occurs by demethylation [57]. It may be that more than one metabolic pathway can lead to an MDMA-induced toxic reaction.

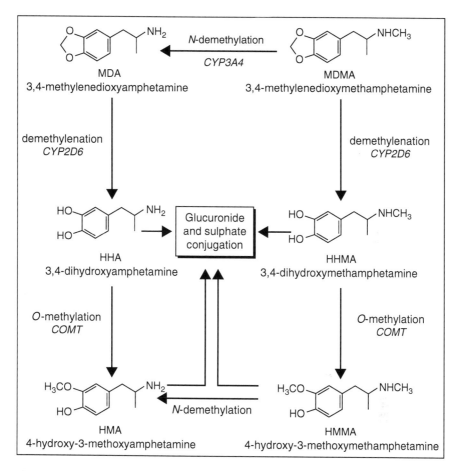

Fig. 10 Main different metabolic pathways of MDMA (Adapted from [55])

Secondary Metabolism of MDMA

A second pathway of MDMA metabolism is N-demethylation to MDA (3,4-methylenedioxyamphetamine), which appears to be but a minor metabolite of MDMA and is an abused drug in its own right (Fig. 10). Concentrations of MDA in plasma range from 3–5% of those corresponding to MDMA [55]. When formed from MDMA, the MDA formation rate constant is approximately 0.75/h and the half-life is 16–28 h, depending on the dose of MDMA given. The Cmax for MDA occurs at 5–7 h, and, on the basis of plasma AUC comparisons of MDMA and MDA, 8–9% of MDMA is converted to MDA, which may be further metabolized before elimination. The urinary recovery of unchanged MDA accounts for approximately 1% of the dose of MDMA [55].

It is unlikely that significant accumulation of MDA would occur after a single dose of MDMA. Given the prolonged half-life of MDA, however, it could accumulate in an individual taking MDMA three or more times/week.

In humans, approximately 50–70% of the total MDMA dose is recovered unchanged in the urine together with some other metabolites [51, 55]. A report based on one patient indicated that after 50 mg of oral, MDMA 32.5 mg (65%) of unchanged drug was excreted in the urine over 72 h [51]. In another study, urine collection showed an increase in the amount of unchanged MDMA excreted by a factor of 20, from the 50-mg to the 150-mg dose, whereas the urinary recovery of 4-hydroxy-3-methoxy-methamphetamine (HMMA) remained unchanged [55]. No significant changes in the urinary pH or creatinine clearance occurred during this study. Although the renal clearance remained fairly constant, the nonrenal clearance appeared to be dose dependent.

Interactions and Overdose with MDMA

MDMA causes a modest increase in blood pressure and pulse rate, in most people, similar to moderate exercise. Because of this, and because a few people may have a more pronounced cardiac response to MDMA, people with a history of high blood pressure, heart trouble, or stroke are advised not to use MDMA, or at the very least are advised to start with a much lower than average dose. The same warning applies to people who are hypersensitive to drugs. Liver or kidney problems are also a contraindication for MDMA use. Deaths have been reported of MDMA users who were also taking Monoamine Oxidase Inhibitors (MAOI's) being prescribed as antidepressants. Therefore MDMA is not recommended to anyone taking any MAOI. In addition, one should be aware that some antidepressants (e.g. Prozac® and Zoloft®) may inhibit some of the effects of MDMA. Although MDMA is thought by many to be a fairly safe drug, it is because of the adulterants, which range from talcum powder to active ingredients such as methamphetamine. Also, "dirty" chemical synthesis in the home labs might contaminate MDMA with by-products often resulting in hallucinations and/or cardiovascular reactions (Table 4).

There is a single report, to our knowledge, of a possible drug interaction involving MDMA and ritonavir, a drug for the treatment of the human immunodeficiency virus [60]. A patient receiving ritonavir ingested MDMA in an estimated dose of 180 mg. The resultant blood MDMA level was 4.56 µg/ml, much higher than would be expected from this dose of MDMA. The authors suggest that the coadministration of MDMA and ritonavir (an inhibitor of CYP2D6) is the explanation for the unusually high levels of MDMA after a commonly used recreational dose. In accord with the notion that CYP2D6 is the responsible enzyme for the main metabolic pathway, preincubation of MDMA with human liver microsomes and nicotinamide adenine dinucleotide phosphate (NADPH) resulted in significant inhibition of CYP2D6 activity. Therefore, MDMA was shown be a potent inhibitor of CYP2D6 in vivo, and the interaction of MDMA with this metabolic pathway may cause long-lasting drug interactions with other CYP2D6 substrates [61]. However, no clinical data are available in support of this theory.

Table 4 Some of the adulterants commonly used in MDMA tablets sold in the street, with concentrations ranging from traces of the active compound to 80% purity of active MDMA (Adapted from [59])

Active ingredients	Side effects	Potential harm
Heroin	CNS depression	Respiratory failure
Amphetamine	CNS activation	Hypertonus
Methamphetamine	CNS overstimulation	Ventric.fiibrillation/Hypertonus
LSD	Hallucination	Paranoia
Ephedrine	CNS activation	Sleeplessness
Caffeine	CNS activation	Restlessness
Paracetamol	Analgesia	Liver necrosis
Acetyl salicylic acid	Analgesia	Gastric ulcers
Strychnine	Central stimulant	Epileptic seizures
Valium, temazepam	Central depressant	Respiratory failure
2,4-Dichloracetic acid, a herbicide contaminated with dioxine	Tachycardia	Cancer
Buflomedil, a vasoactive agent	Hypotonus	
Scopolamine, a muscarinic antagonist highly toxic	Tachycardia, dry mouth, Rubiosis Nausea, Emesis	Epileptic seizures Delirium, Paralysis, Stupor, Death
Atropine	Tachycardia	Hallucination
Meta-chlorphenyl-Piperazin, a serotonin agonist	Headache	Depression
Inert ingredients		
Flour		
Glucose		
Ascorbic acid		
Lactulose		
Borax		
Talcum powder		

Acute Intoxication with MDMA

One of the dangers of MDMA is the apparent lack of relationship between alleged dose and severity of acute toxic reaction with an overlap between serotonin syndrome and neuroleptic malignant syndrome [62, 63]. Although one person attempted suicide after reportedly taking 42 pills with a resultant plasma MDMA level of 7.72 μg/ml and displayed only hypertension and tachycardia [64], others have died with much lower plasma MDMA levels ranging from 0.05–1.26 μg/ml [62]. Furthermore, serum MDMA levels do not correlate well with clinical symptoms [65]. Acute toxic reactions usually develop within 15 min to 6 h after ingestion of MDMA [63]. Symptoms of an acute MDMA toxic reaction include agitation, tachycardia, hypertension, dilated pupils, trismus, and sweating, whereas the more severe cases may be characterized by hyperthermia, disseminated intravascular coagulation (DIC), rhabdomyolysis, and acute renal failure [64]. In more severe cases, elevated creatine kinase levels are often present, [66–69], with levels as high as 12,200–555,000 IU/l being reported [68, 69]. Other frequently reported acute adverse effects occurring after the ingestion of MDMA include lack of appetite, difficulty concentrating, impaired balance, and restless legs [70].

> It is still unclear if the toxic effects after use of MDMA can solely be attributed to the pure agent and/or are due to the impurities during synthesis or the contamination with adulterants such as methamphetamine, cocaine, strychnine, heroin etc.

The toxic effects of MDMA can be divided into three categories to help distinguish acute toxic reactions from long-term residual effects. These categories are

1. Acute reactions at therapeutic doses
2. Overdose reactions and
3. Residual effects

At moderate doses (85–100 mg), acute effects include transient nausea occurring about 30 min after ingestion and lasting about 30 min, increases in both blood pressure and heart rate, and symptoms related to increased muscle tonicity, such as jaw clenching and teeth grinding. In those subjects who are particularly sensitive to MDMA, higher doses (100 mg) cause numbness and tingling in the extremities, luminescence of objects, increased sensitivity to cold, increased color acuity, and vomiting. Residual effects occurring from 2 h to 2 weeks after ingestion included exhaustion, fatigue, and nausea. Doses higher than 200 mg result in a classic toxic psychosis with symptoms of paranoia and auditory and visual hallucinations.

Hyperthermia

Hyperthermia (temperature > 40°C) is the most common adverse effect associated with a severe acute toxic reaction to MDMA. The increase in body temperature is probably due to serotonergic actions of MDMA in the thermo-regulatory center in the hypothalamus [71], because animal studies have shown that the hyperthermia caused by compounds such as MDMA is mediated by actions at serotonin receptors in the CNS [72]. Hyperthermia may also be caused by excessive heat production due to sustained muscle hyperactivity, increased metabolic rate, rigidity, and seizures [73]. Hyperthermia is believed to be the beginning of the cascade leading to disseminated intravascular coagulation (DIC), rhabdomyolysis, myoglobinuria, and acute renal failure. However, the exact pathophysiology of this cascade after MDMA intoxication has not been fully elucidated.

Cardiovascular Effects

Similar to cocaine and amphetamine, acute toxic effects of MDMA may include sympathetic stimulation and increase myocardial oxygen demand, leading to varying degrees of tachycardia, vasoconstriction, changes in blood pressure, and arrhythmias. In severe cases, vasospasm leading to acute myocardial infarction and irreversible dilated cardiomyopathy may occur [74]. Abnormal electrocardiographic changes that show widespread ST-segment elevation indicating acute myocardial infarction have been in MDMA users [75]. During postmortem evaluations, necrosis of the heart (contraction band necrosis or widespread foci of necrosis) has been seen [76] and may be due to excessive catecholamine release [74]. These findings do not necessarily establish a cause and effect relationship, since other substances or circumstances (coadministration of drugs) may have contributed to these observed cardiovascular effects.

Cerebrovascular Effects

Use of MDMA can be associated with intracerebral hemorrhage often in conjunction with an underlying vascular malformation [77]. Other investigators have postulated that those who take MDMA are at an increased risk for cerebrovascular accidents due to the altered 5-HT system, because postsynaptic 5-HT receptors are involved in the regulation of the brain microvasculature [78]. Other cerebrovascular adverse effects that have been associated with MDMA include subarachnoid hemorrhage, cerebral infarction, and cerebral venous sinus thrombosis [79]. Magnetic resonance imaging even revealed a left basal ganglia hematoma after the ingestion of MDMA in a patient with no apparent cardiovascular risk factors [80].

Neuroendocrine Effects

Numerous cases of hyponatremia have been associated with MDMA use, often in combination with seizures, catatonic stupor, and incontinence of urine [81–84]. It is possible that hyponatremia is a direct result of MDMA neuroendocrine effects or from massive water intake leading to dilutional hyponatremia. Since many users take MDMA during all-night dancing parties, large amounts of fluid are ingested, both as a natural consequence of physical activity and because of MDMA-induced hyperthermia. Hyponatremia may be due to the syndrome of inappropriate antidiuretic hormone [84] because MDMA causes the release of arginine vasopressin (AVP) [85]. In addition, the extreme dehydration caused by sweating and/or vomiting associated with MDMA use combined with massive water intake could lead to hyponatremia [86]. In one report of a fatality due to MDMA, hyponatremia leading to cerebral edema appeared to be the main cause of death [87]. Most likely, contamination of the MDMA tablets with other substances has been postulated as the cause of hyponatremia associated with MDMA use [82]. Postcards have been distributed in some clubs and bars advising patrons who take MDMA that they should drink about a pint of water an hour and eat or drink something salty, such as a sports drink, to replace lost sodium [88]. Also, MDMA and its metabolites can stimulate both oxytocin and vasopressin release in vitro, the response being dose dependent for each drug with hydroxymethoxy mandelic acid (HMMA) being the most potent [89].

Hepatotoxicity

Hepatotoxic effects have been associated with MDMA [62, 76, 90–92]. In one series of seven fatalities associated with the use of ring-substituted amphetamines, including MDMA, necrosis of the liver was seen in all cases [76]. Liver transplantation had to be performed [91]. Two of the most likely mechanisms for causing a hepatotoxic reaction are immune-mediated reaction or injury secondary to hyperthermia [92]. However, a hepatotoxic reaction arising from drug impurities or MDMA metabolites may also be a likely explanation.

Psychopathology

A psychotic syndrome characterized by delusions, usually of the persecutory type, may be caused by MDMA. Other nonpsychotic conditions include visual phenomena, depersonalization and derealization, panic attacks, and depression. Persons who display such symptoms may have at least one first-degree relative with a history of psychiatric illness and be predisposed to have psychiatric symptoms [93]. Anxiety attacks, persistent insomnia, rage reactions, and psychosis (especially at higher doses) have occurred after MDMA use, although in most cases the psychiatric status of these patients was not known [94]. Compared with control subjects who do not take MDMA, those who frequently take MDMA have significantly higher scores on psychiatric scales used to assess somatization, obsession, anxiety, hostility, phobia psychosis, paranoid ideas, psychoses, poor appetite, and restless or disturbed sleep, also showing greater impulsiveness [95].

Treatment of MDMA-Related Toxicity

The diagnosis of acute toxic reaction to MDMA is based on the history and clinical features of intoxication. Initial examination should include blood chemistry analysis, complete blood count, liver function tests, cardiac enzyme and creatinine kinase

> **There is no antidote for MDMA intoxication. Recommended treatment of MDMA overdose is similar to the treatment of cocaine or methamphetamine overdose. The first priority should be maintaining the airway, breathing, and circulation.**

measurements, and a urine toxicology screen. Quantitative serum levels do not correlate well with severity of symptoms and are not generally available [96]. A complete history and physical examination should be performed, and the patient should be assessed for hypertensive crisis or life-threatening arrhythmias. An electrocardiogram for chest pain or a computed tomographic (CT) scan of the brain for persistent mental status changes should be obtained [97, 98]. Amphetamines and related drugs (i.e., methamphetamine, MDMA) can be detected in the urine, but there is a high degree of cross-reactivity between amphetamine derivatives and adrenergic amines. Therefore, confirmatory testing usually is required [65].

Thereafter, treatment will be aimed at reducing various symptoms, including hyperthermia, agitation, cardiovascular and cerebrovascular incidents, neuroendocrine abnormalities, and neurologic problems.

First, lavage of the gastrointestinal tract, activated charcoal, and cathartic techniques is advocated. Induction of emesis is not appropriate because of the potential for CNS depression and seizures [65]. Because approximately 50–70% of MDMA is recovered in the urine [55] [51], renal failure would significantly decrease the elimination of MDMA from the body, so maintaining adequate hydration is essential. Because MDMA is a weak base and a significant proportion is eliminated in the urine, acidifying the urine is likely to be an effective means of increasing renal elimination, but it may precipitate acute renal failure in patients with myoglobinuria and is not recommended [65].

Hyperthermia

Although fatalities may be due to many different causes, hyperthermia is probably the single most important condition to treat because it may lead to further severe complications, such as rhabdomyolysis and disseminated intravascular coagulation (DIC) [99]. Since mortality has been correlated to both the extent of hyperthermia and the duration, aggressive active cooling measures are indicated in cases of MDMA-induced hyperthermia [71]. It is important to control agitation to limit further heat production [100]. Neuromuscular blockers, such as pancuronium, have been given, but their use requires ventilation and endotracheal intubation [65].

Dantrolene sodium, a drug that is indicated for the treatment of malignant hyperthermia and that inhibits the release of calcium from the sarcoplasmic reticulum, is recommended by many clinicians to treat hyperthermia secondary to MDMA use [101–105]. Speculative hypotheses notwithstanding, the use of dantrolene for the treatment of MDMA intoxication remains controversial. The efficacy of dantrolene in treating this condition has been questioned, as some patients have improved with supportive care only [106] and some clinicians assert that there is insufficient evidence to recommend dantrolene in cases of MDMA acute toxic reaction [107–109].

To determine if MDMA caused an increase of calcium within the muscle, which would suggest that an inhibitor of calcium release in skeletal muscle, such as dantrolene, might be efficacious in treating MDMA intoxication, in vitro experiments using human muscle subjected to halothane and caffeine contracture tests were performed (used to test for susceptibility to malignant hyperthermia). The results indicated that the hyperthermia from MDMA intoxication is associated with an elevation in the myoplasmic calcium concentrations, similar to that seen in malignant hyperthermia, which suggests that dantrolene might be a helpful agent in treating MDMA-induced hyperthermia. It has been argued that MDMA-induced hyperthermia results from augmentation of central serotonin, and since dantrolene has no central activity (inhibits calcium peripherally in the skeletal muscle), it should not be effective. Therefore, a non-depolarizing neuromuscular blocker may be just as effective in treating MDMA acute toxic reaction [63]. Although, there are not sufficient data in humans to confirm that the hyperpyrexia associated with MDMA is a centrally mediated effect, the use of dantrolene should not be precluded because it does appear to reduce pyrexia secondary to exertional heatstroke [104]. It is also hypothesized that the unpredictable hyperthermia associated with MDMA may result from an underlying metabolic myopathy, similar to that seen with exertional heatstroke, and associated with a skeletal muscle abnormality similar to malignant hyperthermia [110].

Cardiovascular Treatment

Tachycardia without hemodynamic compromise does not need to be treated. Sedative dosages of benzodiazepines may be helpful by reducing blood pressure and heart rate, which may reduce myocardial oxygen demand [111]. β-Blockers such as propranolol should be avoided when treating stimulant-induced hypertension because this may result in unopposed α-adrenergic vasoconstriction. Hypertension can be treated with an α-blocker such as phentolamine or with a direct-acting vasodilator such as nitroprusside [65, 74]. Another option is the use of a β-blocker concurrently with phentolamine [100]. Myocardial ischemia caused by stimulants should be treated with oxygen, aspirin, and benzodiazepines. If these options do not reverse the ischemia, then vasodilators or phentolamine should be given [74]. Arterial spasm may be treated with sublingual or intravenous nitroglycerin [65]. Arrhythmias should be treated accordingly with a short acting and selective ß-blocker like esmolol (Brevibloc®), which due to its short half-life can be administered to effect. Thrombolytic agents can be taken into consideration as they have been given safely to patients with stimulant-induced myocardial infarction [111].

Cerebrovascular Treatment

Patients with altered mental status, lethargy, or obtundation should undergo CT of the brain because of the risk for intracranial hemorrhage and infarct [97]. In patients with nontraumatic intracranial hemorrhage, arteriography should be performed and a thorough history of the use of illicit substances should be evaluated [77].

Neurologic Treatment

Patients who are agitated may require treatment with a benzodiazepine, such as diazepam, lorazepam, or midazolam [65, 97]. It is very important to control agitation in order to decrease further heat production [100]. Some of the conditions associated with MDMA, namely acute toxic reactions (mental status changes, hyperthermia, autonomic instability, increased motor restlessness, myoclonus, elevated creatine kinase level, diaphoresis, and death due to renal failure) are similar to the findings in both neuroleptic malignant syndrome and serotonin syndrome [62, 101]. Pharmacologic treatments effective in these syndromes are recommended by some clinicians [101] including methysergide maleate, a nonspecific serotonin antagonist [112], ß-blockers, 5-HT1_A antagonists [113], or bromocriptine, a dopamine agonist [114]. However, none of these drugs has been prospectively evaluated for the treatment of MDMA acute toxic reaction.

Caution may be warranted in using antipsychotic agents when treating MDMA intoxication. Antipsychotics decrease the seizure threshold, and blocking dopamine receptors may affect the thermoregulatory system leading to hyperthermia or exacerbation of existing hyperthermia. In addition, SSRIs may further increase serotonergic transmission by blocking the reuptake of synaptic serotonin, possibly raising the risk for development of the serotonin syndrome or aggravating already existing hyperthermia [115].

Owing to the risk for hepatotoxicity, it would be prudent to monitor liver function in persons suspected of taking MDMA [92] and any person with unexplained jaundice or hepatomegaly should be screened for a history of MDMA use or other stimulating drugs [90]. Treatment will be primarily supportive. If severe hepatic necrosis has occurred, transplantation may be the only option, and has been performed successfully in patients with acute liver failure due to MDMA use [91, 73].

In summary, supportive care in individuals presenting with MDMA intoxication primarily includes rehydration with intravenous fluids and lowering the temperature of the patient with use of cooling blankets or ice baths [114]. In some cases, lowering the body temperature may require infusion of cold intravenous fluids or peritoneal lavage with cool dialysate [63]. Crystalloids may be given to help treat both the profuse sweating that often accompanies MDMA acute toxic reaction as well as prophylaxis against acute renal failure secondary to rhabdomyolysis and myoglobinuria [66]. Furthermore, judicious fluid support may help with symptoms of hepatotoxicity as it may increase liver blood flow and prevent further hepatic damage [92]. The following the steps outlined below should be pursued if there is a remote suspicion of an amphetamine-related intoxication (Table 5).

Table 5 Summary of management in MDMA related toxicity

Prevent further absorption

Apply activated charcoal 50 g p.o. if <1 h post-ingestion

Monitor

For at least 4 h pulse, blood pressure, ECG, core temperature

Check

Blood urea, electrolytes, creatinine, liver function, CPK

Consider clotting profile and arterial blood gases

Urine drug screen(UDS)

Positive result for methamphetamine helps to confirm consumption with specific tests available (see chapter on Urine Drug Screen; page 251)

Treat

Anxiety or agitation – diazepam (0.1–0.3 mg kg^{-1}) p.o. or i.v.

Seizures – diazepam (0.1–0.3 mg kg^{-1}) i.v. or per rectum (p.r.)

Hyponatraemia – fluid restrict, consider hypertonic saline if severe

Metabolic acidosis – correct, if QT interval is prolonged, by using sodium bicarbonate

Severe hypertension – consider labetalol

Hypotension – intravascular volume expansion, consider need for central venous access, cardiac output monitoring, etc.

Hyperthermia – use simple cooling methods. If initial temperature >39°C, give dantrolene; intubation and ventilation are likely to be required

Organ-system failure – conventional support; promote diuresis of 1–2 ml kg^{-1} h^{-1} with mannitol or furosemide

Appendix
Analogues of MDMA and Related Compounds

MDMA has several chemical "cousins" which have different effects. Briefly, here are descriptions of some of the more common ones [4, 116].

MDA (3,4-methylenedioxyamphetamine)

MDA (Fig. 11) was popular for a while during the 1970s, when it was known as the 'Love Drug', a nickname sometimes associated with MDMA as well. It is similar to MDMA in its effects, but is slightly more stimulating. It has been shown in laboratory studies to be approximately twice as neurotoxic as MDMA, though in some 30 years of human use, case reports do not suggest that it has caused behavioral or psychological problems.

MDE or MDEA (N-ethyl-methylenedioxyamphetamine)

Commonly called "Eve" (if MDMA is "Adam", MDE is "Eve"; Fig. 12) MDE is similar to MDMA, though it seems to turn the subject inwards according to some users it invites less communication than does MDMA.

MMDA (3-methoxy-4,5-methylenedioxyamphetamine)

Often confused with the similarly-named but chemically different MDMA. MMDA (Fig. 13) is reported to generate interesting, closed-eye hallucinations, i.e. "brain movies", or conscious dreams.

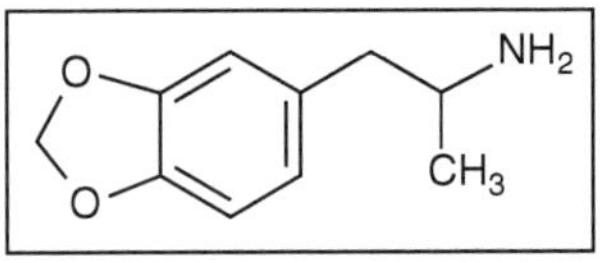

Fig. 11 Molecular structure of MDA, a cousin of MDMA

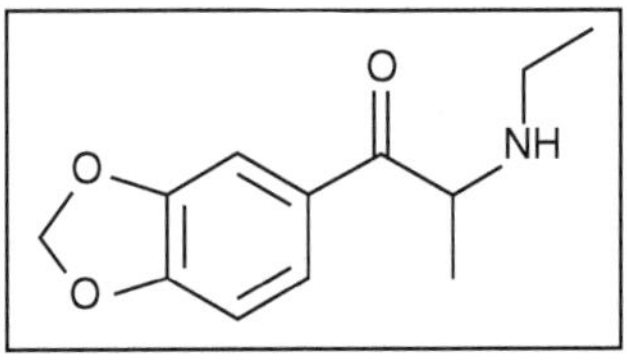

Fig. 12 Molecular structure of MDEA, the sister of MDMA

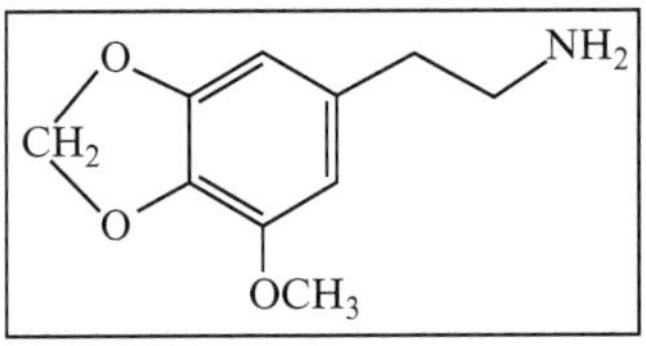

Fig. 13 Molecular structure of MMA, a relative of MDA with psychoactive properties

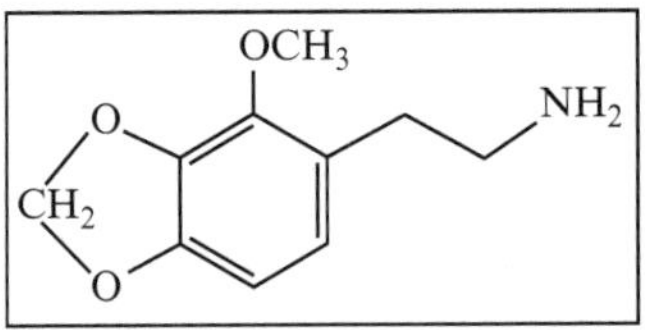

Fig. 14 The isomer of MMA, which in contrast has a ethylamine side-chain relocated to a position adjacent to the methoxyl group

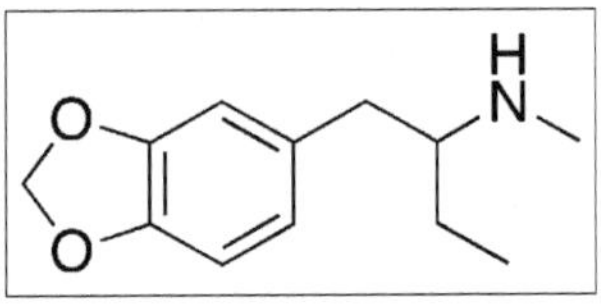

Fig. 15 MBDB (Eden or Methyl-J) is a synthetic empathogen with effects similar to but milder than those of MDMA. It is very uncommon

2-Methoxy-3,4-methylenedioxyphenethylamine

This is another close relative of MMA (Fig. 14). It induces a pleasant mood elevation with no psychotomimetic effects.

MBDB (N-methyl-1-(1,3-benzodioxol-5-yl)-N-methylbutan-2-amine)

This MDMA derivative differs structurally only by the addition of an extra carbon to the MDMA chain (Fig. 15). The effects are similar to MDA.

References

1. Freudenmann RW, Öxler F, Bernschneider-Reif S. The origin of MDMA (ecstasy) revisited: the true story reconstructed from the original documents. Addiction. 2006;101:1241–1245.
2. Benzenhöfer U, Passie T. The early history of "Ecstasy." (Article in German) Nervenarzt. 2006;77:95–99.
3. Shulgin A, Nichols D. Characterization of three new psychomimetics. In: Stillmann R, Wilette R, editors. The Psychopharmacology of Halluzinogens. New York: Pergamon; 1978. pp. 74–82.
4. Shulgin A. Pihkal: a chemical love story. Berkeley: Transform Press; 1995.
5. Greer G. Using MDMA in psychotherapy. Advances. 1985;2:57–59.
6. McGlothlin WH, Arnold DO. LSD revisited: a ten-year follow-up of the medical LSD use. Arch Gen Psychiatr. 1971;25:35–49.
7. Naranjo C. MDMA: the drug of analysis. New York: Ballantine Books; 1973.
8. Abramson HA. The use of LSD in psychotherapy and alcoholism. New York: Bobbs-Merrill; 1967.
9. Shulgin AT, Carter MF. Centrally active phenethylamines. Psychopharm Commun. 1974;1:93–98.
10. Margolis J. Ecstasy's dividend. Time. 2001;19:58–59.
11. Schmidt WJ, Mayerhofer A, Meyera A, Kovar KA. Ecstasy counteracts catalepsy in rats, an anti-parkinsonian effect? Neurosci Lett. 2002;330:251–254.
12. Nair SG, Gudelsky GA. Protein kinase C inhibition differentially affects 3,4-methylenedioxymethamphetamine-induced dopamine release in the striatum and prefrontal cortex of the rat. Brain Res. 2004;1013:168–173.
13. Homberg J, de Boer S, Raaso HS, Olivier JDA, Verheul M, Ronken E, et al. Adaptations in pre- and postsynaptic 5-HT1A receptor function and cocaine supersensitivity in serotonin transporter knockout rats. Psychopharmacology. 2008;200:367–380.
14. Mendelson J, editor. The Pharmacology of Ecstasy. MDMA/Ecstasy Research: Advances, Challenges and Future Directions; 2001; Bethesda, MD: US Goverment.
15. State RC, Grob CS, Poland RE, editors. Psychobiologic Effects of 3,4-Methylenedioxymethamphetamine in Humans: A Pilot Study (Poster). MDMA/Ecstasy Research: Advances, Challenges and Future Directions; 2001; Bethesda, MD: US Goverment.
16. Gamma A, Buck A, Berthold T, Hell D, Vollenweider FX. 3,4-Methylenedioxymethamphetamine (MDMA) modulates cortical and limbic brain activity as measured by [H215O]-PET in healthy human. Neuropsychopharmacology. 2000;23:388–395.
17. Green AR, Mechan AO, Melliot JM, O'shea E, Colado MI. The pharmacology and clinical pharmacology of 3,4-methylenedioxymethamphetamine (MDMA, "Ecstasy"). Pharmacol Rev. 2003;55:463–508.
18. Bankson MG, Cunningham KA. 3,4-Methylenedioxymethamphetamine (MDMA) as a unique model of serotonin receptor function and serotonin-dopamine interactions. J Pharmacol Exp Ther. 2001;297:846–852.

19. Reneman L, Booij J, Schmand B, van den Brink W, Gunning B. Memory disturbances in "Ecstasy" users are correlated with an altered brain serotonin neurotransmission. Psychopharmacology. 2000;148:322–424.
20. Liechti ME, Geyer MA, Hell ID, Vollenweider FX. Effects of MDMA (Ecstasy) on prepulse inhibition and habituation of startle in humans after pretreatment with citalopram, haloperidol, or ketanserin. Neuropsychopharmacology. 2001;24:240–252.
21. Liester MB, Grob CS, Bravo GL, Walsh RN. Phenomenology and sequelae of 3,4-methylenedioxymethamphetamine use. J Nerv Ment Dis. 1992;180:345–352.
22. Ricaurte G, Bryan G, Strauss L, Seiden L, Schuster C. Hallucinogenic amphetamine selectively destroys brain serotonin nerve terminals: neurochemical and anatomical evidence. Science. 1985;229:986–988.
23. Lamers GTJ, Ramaekers JG, Muntjewerff ND, Sikkema KL, Riedel WJ, Samyn N, et al. Dissociable effects of a single dose of ecstasy (MDMA) on psychomotor skills and attentional performance. J Psychopharm. 2003;17:379–387.
24. Mathias R. "Ecstasy" damages the brain and impairs memory in humans. NIDA Notes. 1999;14:10–15.
25. Lester SJ, Baggott M, Welm S, Schiller NB, Jones RT, Foster E, et al. Cardiovascular effects of 3,4-methylenedioxymethamphetamine: a double-blind, placebo-controlled trial. Ann Intern Med. 2000;133:969–972.
26. Grob CS, Poland RE, Chang L, Ernst T. Psychobiologic effects of 3,4-methylenedioxymethamphetamine in humans: methodological considerations and preliminary observations. Behav Brain Res. 1996;73:103–107.
27. Hall AP, Henry JA. Acute toxic effects of 'Ecstasy' (MDMA) and related compounds: overview of pathophysiology and clinical management. Br J Anaesth. 2006;96:678–685.
28. Davis WM, Hatoum HT, Waters IW. Toxicity of MDA (3,4-methylenedioxyamphetamine) considered for relevance to hazards of MDMA (Ecstasy) abuse. Alcohol Drug Res.. 1987;7:123–134.
29. Peroutka SJ, Pascoe N, Faull KF. Monoamine metabolites in the cerebrospinal fluid of recreational users of 3 4-methylenedioxymethamphetamine (MDMA, 'Ecstasy'). Res Commun Subst Abuse. 1987;8:125–138.
30. Ricaurte GA, Forno LS, Wilson MA, DeLanney L, Irwin I, Molliver ME, et al. (±)3,4-Methylenedioxymethamphetamine selectively damages central serotonergic neurons in nonhuman primates. JAMA. 1988;260:51–55.
31. Ricaurte GA, Matello AL, Katz JL, Martello MB. Long lasting effects of (±)-3,4-methylenedioxyamphetamine (MDMA) on central serotonergic neurons in nonhuman primates: neurochemical observations. J Pharmacol Exp Ther. 1992;261:616–622.
32. Reneman L, Booij J, den Heeten GJ, van den Brin W. Effects of MDMA (ecstasy) use and abstention on serotonin neurons. Lancet. 2002;359:1617–1618.
33. Ricaurte GA, Yuan J, Hatzidimitriou G, Cord BJ, McCann UD. Severe dopaminergic neurotoxicity in primates after a common recreational dose regimen of MDMA ("Ecstasy"). Science. 2002;297:2260–2263.
34. Grob CS. Deconstructing ecstasy: the epolitics of MDMA research. Addict Res. 2000;6:549–588.
35. Reneman L, Booij J, Lavalaye J, de Bruin K, Reitsma J, Gunning B, et al. Use of amphetamine by recreational users of ecstasy (MDMA) is associated with reduced striatal dopamine transporter densities: a [123I]beta-CIT SPECT study - preliminary report. Psychopharmacology. 2002;159:335–340.
36. Sotnikova TD, Beaulieu JM, Barak LS, Wetsel WC, Caron MG. Dopamine-independent locomotor actions of amphetamines in a novel acute mouse model of parkinson disease. PLoS Biol. 2005;3:e271.
37. Morris K. Concern over research reawakens ecstasy neurotoxicity debate. Lancet. 2003;2:650.
38. Buchert R, Thomasius R, Nebeling B, Petersen K, Obrocki J, Jenicke L, et al. Long-term effects of "Ecstasy" use on serotonin transporters of the brain investigated by PET. J Nucl Med. 2003;44:375–384.

39. Hatzidimitriou G, McCann UD, Ricaurte GA. Altered serotonin innervation patterns in the forebrain of monkeys treated with MDMA seven years previously: factors influencing abnormal recovery. J Neurosci. 1999;19:5096–5107.
40. Hiramatsu M, Kumagai Y, Unger SE, Cho AK. Metabolism of methylenedioxymethamphetamine: formation of dihydroxy-methamphetamine and a quinone identified as its glutathione adduct. J Pharmacol Exp Ther. 1990;25:521–527.
41. Tucker GT, Lennard MS, Ellis SW, et al. The demethylenation of methylenedioxymethamphetamine ("Ecstasy") by debrisoquine hydroxylase (CYP2D6). Biochem Pharmacol. 1994;47: 1151–1156.
42. Zhao ZY, Castagnoli NJ, Ricaurte GA, Steele T, Martello M. Synthesis and neurotoxicological evaluation of putative metabolites of the serotonergic neurotoxin 2-(methylamino)-1-[3,4-(methylenedioxy)phenyl] propane [(methylenedioxy)-methamphetamine]. Chem Res Toxocol. 1992;5:89–94.
43. Lim HK, Foltz RL. In vivo and in vitro metabolism of 3,4-(methylenedioxy)methamphetamine in the rat: identification of metabolites using an ion trap detector. Chem Res Toxicol. 1988;1: 370–378.
44. Wagner GC, Carelli RM, Jarvis MF. Ascorbic acid reduces the dopamine depletion induced by methamphetamine and the 1-methyl-4-phenyl pyridinium ion. Neuropharmacology. 1086;25: 559–561.
45. Bindoli A, Rigobello MP, Deeble DJ. Biochemical and toxicological properties of the oxidation products of catecholamines. Free Radical Biol Med. 1992;13:391–405.
46. Wagner G, Carelli R, Jarvis M. Pretreatment with ascorbic acid attenuates the neurotoxic effects of methamphetamine in rats. Res Commun Chem Path Pharm. 1985;47:221–228.
47. Leibovitz B. Phenethylamines, free radicals, and antioxidants. Multidiscip Ass Psychodel Stud. 1993;4:1–3.
48. McCann UD, Ricaurte CA. Reinforcing subjective effects of (±) 3,4-methylenedioxymethaniphetamine ("ecstasy") may be separable from its neurotoxic actions: clinical evidence. J Clin Psychopharm. 1993;13:214–217.
49. Mas M, Farre M, de la Torre R, et al. Cardiovascular and neuroendocrine effects and pharmacokinetics of 3,4-methylenedioxymethamphetamine in humans. J Pharmacol Exp Ther. 1999;290:136–145.
50. Cami J, de la Torre R, Ortuno J, et al. Pharmacokinetics of ecstasy (MDMA) in healthy subjects [abstr]. Eur J Clin Pharmacol. 1997;52:A168.
51. Verebey K, Alrazi J, Jaffee JH. The complications of 'Ecstasy' (MDMA). JAMA. 1988;259: 1649–1650.
52. Helmlin HJ, Bracher K, Bourquin D, Vonlanthen D, Brenneisen R. Analysis of 3,4-methylenedioxymeth-amphetamine (MDMA) and its metabolites in plasma and urine by HPLC-DAD and GC-MS. J Anal Toxicol. 1996;20:432–440.
53. Hiramatsu M, DiStefano EW, Cho AK. Stereochemical differences in the in vivo and in vitro metabolism of MDMA [abstr]. FASEB. 1989;3:A1035.
54. Lin LY, Kumagai Y, Cho AK. Enzymatic and chemical demethylenation of (methylenedioxy) amphetamine and (methylenedioxy)methamphetamine by rat brain microsomes. Chem Res Toxicol. 1992;5:401–406.
55. de la Torre R, Farré M, Ortuno J, Mas M, Brenneisen R, Roset PN, et al. Non-linear pharmacokinetics of MDMA ('ecstasy') in humans. Br J Clin Pharmacol. 2000;49:104–109.
56. Lin LY, DiStefano EW, Schmitz DA, et al. Oxidation of methamphetamine and methylenedioxymethamphetamine by CYP2D6. Drug Metab Dispos. 1997;25:1059–1064.
57. Kreth KP, Kovar KA, Schwab M, Zanger UM. Identification of the human cytochrome P450 involved in the oxidative metabolism of "Ecstasy"-related designer drugs. Biochem Pharmacol. 2000;59:1563–1571.
58. Schwab M, Seyringe RE, Brauer RB, Hellinger A, Griese EU. Fatal MDMA intoxication. Lancet. 1999;353:593–594.
59. Saunders N. Ecstasy. In: Walder P, editor. Ecstasy und verwandte Designerdrogen. Zürich: Ricco Bilger; 1994. pp. 26–42.

60. Henry JA, Hill IR. Fatal interaction between ritonavir and MDMA. Lancet. 1998;352: 1751–1752.
61. Kreth K, Kovar K, Schwab M, Zanger UM. Identification of the human cytochromes P450 involved in the oxidative metabolism of "Ecstasy" – related drugs. Biochem Pharmacol. 2000;69:1563–1571.
62. Demirkiran M, Jankovic J, Dean JM. Ecstasy intoxication: an overlap between serotonin syndrome and neuroleptic malignant syndrome. Clin Neuropharmacol. 1996;19:157–164.
63. Dar KJ, McBrien ME. MDMA-induced hyperthermia: report of a fatality and review of current therapy. Inten Care Med. 1996;22:995–996.
64. Henry JA, Jeffreys KJ, Dawling S. Toxicity and deaths from 3,4-methylene-dioxymethamphetamine. Lancet. 1992;340:384–387.
65. Benowitz NL. Amphetamines. In: Olson KR, editor. Poisoning and drug overdose. Stamford, CT: Appleton & Lange; 1999. pp. 68–70.
66. Singarajah C, Lavies NG. An overdose of ecstasy: a role for dantrolene. Anaesthesia. 1992;47: 686–687.
67. Mallick A, Bodenham AR. MDMA-induced hyperthermia: a survivor with an initial body temperature of 42.9°C. J Accid Emerg Med. 1997;14:336–338.
68. Murthy BVS, Wilkes RG, Roberts NB. Creatine kinase isoform changes following ecstasy overdose. Anaesth Inten Care. 1997;25:156–159.
69. Hall AP, Lyburn ID, Spears FD, Riley B. An unusual case of Ecstasy poisoning. Inten Care Med. 1996;22:670–671.
70. Vollenweider FX, Gamma A, Liechti M, Huber T. Psychological and cardiovascular effects and short-term sequelae of MDMA ("Ecstasy") in MDMA-naïve healthy volunteers. Neuropsychopharmacology. 1998;19:241–251.
71. Hall AP. "Ecstasy" and the anaesthetist. Br J Anaesth. 1997;79:697–698.
72. Schmidt CJ, Black CK, Abbate GM, Taylor VL. MDMA-induced hyperthermia and neurotoxicity are independently mediated by 5-HT2 receptors. Brain Res. 1990;529:85–90.
73. Olson KR. Comprehensive evaluation and treatment. In: Olson KR, editor. Poisoning and drug overdose. Stamford, CT: Appleton & Lange; 1999. pp. 1–61.
74. Ghuran A, Nolan J. Recreational drug misuse: issues for the cardiologist. Heart. 2000;83: 627–633.
75. Qasim A, Townend J, Davies MK. Ecstasy induced acute myocardial infarction. Heart. 2001;85:E10.
76. Milroy CM, Clark JC, Forrest ARW. Pathology of deaths associated with "Ecstasy" and "Eve" misuse. J Clin Pathol. 1996;49:149–153.
77. McEvoy AW, Kitchen ND, Thomas DG. Intracerebral haemorrhage and drug abuse in young adults. Br J Neurosurg. 2000;14:449–454.
78. Reneman L, Habraken JB, Majoie CB, Booij J, den Heeten GJ. MDMA ("Ecstasy") and its association with cerebrovascular accidents: preliminary findings. Am J Neuroradiol. 2002;21: 1001–1007.
79. McCann U, Slate SO, Ricaurte GA. Adverse reactions with 3,4-methylenedioxymethamphetamine (MDMA, 'Ecstasy'). Drug Saf. 1996;15:107–115.
80. Ranalli E, Bouton R. Intracerebral haemorrhage associated with ingestion of "Ecstasy" [abstr]. Eur Neuropsychopharmacol. 1997;7:S263.
81. Maxwell DL, Polkey MI, Henry JA. Hyponatraemia and catatonic stupor after taking "Ecstasy". BMJ. 1993;307:1399.
82. Kessel B. Hyponatraemia after ingestion of "Ecstasy" BMJ. 1994;308:414.
83. Holden R, Jackson MA. Near-fatal hyponatraemic coma due to vasopressin over-secretion after "Ecstasy" (3,4-MDMA) [letter]. Lancet. 1996;347:1052.
84. Matthai SM, Davidson DC. Cerebral oedema after ingestion of MDMA ("Ecstasy") and unrestricted intake of water [letter]. BMJ. 1996;312:1359.
85. Henry JA, Fallon JK, Kicman AT, et al. Low-dose MDMA ("Ecstasy") induces vasopressin secretion [letter]. Lancet. 1998;351:1784.

86. Wilkins B. Cerebral oedema after MDMA ("Ecstasy") and unrestricted water intake: hyponatraemia must be treated with low water input. BMJ. 1999;313:689–690.
87. Parr MJA, Low HM, Botterill P. Hyponatremia and death after "Ecstasy" ingestion. Med J Aust. 1997;166:136–137.
88. Finch E, Sell L, Arnold D. Cerebral oedema after MDMA ("Ecstasy") and unrestricted water intake: drug workers emphasize that water is not an antidote to drug [letter]. BMJ. 1996;313:690.
89. Forsling ML, Fallon JK, Shah D, Tilbrook GS, Cowan DA, Kicman AT, et al. The effect of 3,4-methylenedioxymethamphetamine (MDMA, 'ecstasy') and its metabolites on neurohypophysial hormone release from the isolated rat hypothalamus. Br J Pharmacol. 2002;135:649–656.
90. Henry JA, Jeffreys KJ, Dawling S. Toxicity and deaths from 3,4-methylenedioxymethamphetamine ("ectasy"). Lancet. 1992;340:384–387.
91. Ellis AJ, Wendon JA, Portmann B, Williams R. Acute liver damage and ecstasy ingestion. Gut. 1996;38:454–458.
92. Jones AL, Simpson KJ. Review article: mechanisms and management of hepatotoxicity in ecstasy (MDMA) and amphetamine intoxications. Aliment Pharmacol Ther. 1999;13:129–133.
93. McGuire PK, Cope H, Fahy TA. Diversity of psychopathology associated with use of 3,4-methylenedioxymethamphetamine ('Ecstasy'). Br J Psychiatr. 1994;165:391–395.
94. Hayner GN, McKinney H. MDMA: the dark side of ecstasy. J Psychoactive Drugs. 1986;18:341–347.
95. Parrott AC, Sisk E, Turner JJ. Psychobiological problems in heavy 'Ecstasy' (MDMA) polydrug users. Drug Alcohol Depend. 2000;60:105–110.
96. Dowling GP, McDonough ET, Bost RO. 'Eve' and 'Ecstasy': a report of five deaths associated with the use of MDEA and MDMA. JAMA. 1987;257:1615–1617.
97. Perrone J. Amphetamines. In: Viccellio P, editor. Emergency toxicology. Philadelphia: Lippincott-Raven; 1998. pp. 899–902.
98. Rochester JA, Kirchner JT. Ecstasy (3,4-methylene-dioxymethamphetamine): history, neurochemistry, and toxicology. J Am Board Fam Pract. 1999;12:137–142.
99. Screaton GR, Singer M, Cairns HS, Thrasher A, Sarne RM, Cohen SL. Hyperpyrexia and rhabdomyolysis after MDMA ("Ecstasy") abuse [letter]. Lancet. 1992;339:677–678.
100. McKinney PE. Designer drugs. In: Haddad LM, Shannon MW, Winchester JF, editors. Poisoning and drug overdose. Philadelphia: WB Saunders; 1998. pp. 569–580.
101. Ames D, Wirshing WC. Ecstasy, the serotonin syndrome, and neuroleptic malignant syndrome – a possible link. J Am Med Assoc. 1992;268:1505–1506.
102. Tehan B. Tehan B. Ecstasy and dantrolene [letter]. BMJ. 1993;306:146.
103. Logan AS, Stickle B, O'Keefe N. Survival following "Ecstasy" ingestion with a peak temperature of 42 degrees. Anaesthesia. 1993;48:1017–1018.
104. Larner AJ. Dantrolene and 'Ecstasy' overdose. Anaesthesia. 1993;38:179–180.
105. Singarajah C, Lavies NG. An overdose of ectasay. A role for dantrolene. Anaesthesia. 1992;47:686–687.
106. Campkin NTA, Davies UM. Treatment of 'Ecstasy' overdose with dantrolene [letter]. Anaesthesia. 1993;48:82–83.
107. Watson JD, Ferguson C, Hinds CJ, Skinner R, Coakley JH. Exceptional heatstroke induced by amphetamine analogues - does dantrolene have a place. Anaesthesia. 1993;48:542–543.
108. Dowsett RP. Deaths attributed to "Ecstasy" overdose [letter]. Med J Aust. 1996;164:700.
109. Chadwick IS, Curry PD, Linsley A, Freemont AJ, Doran B. Ecstasy, 3,4-methylenendioxyamphetamine (MDMA), a fatality associated with coagulopathy and hyperthermia [letter]. J R Soc Med. 1991;84:371.
110. Hopkins PM, Ellis FR, Halsall PJ. Evidence for related myopathies in exertional heat stroke and malignant hyperthermia. Lancet. 1991;338:1491–1492.
111. Hollander JE. The management of cocaine-associated myocardial ischemia. N Engl J Med. 1995;333:1267–1272.
112. Sandyk R. L-dopa induced 'serotonin syndrome' in a parkinsonian patient on bromocriptine. J Clin Psychopharmacol. 1986;6:194–195.

113. Guze BH, Baxter LRJ. The serotonin syndrome: case responsive to propranolol [letter]. J Clin Psychopharmacol. 1986;6:119–120.
114. McDowell DM. MDMA, ketamine, GHB, and the "club drug" scene. In: Galanter M, Kleber HD, editors. Textbook of substance abuse treatment. Washington, DC: American Psychiatric Press; 1999. pp. 295–305.
115. Green AR, Cross AJ, Goodwin GM. Review of the pharmacology and clinical pharmacology of 3,4-methylenedioxymethamphetamine (MDMA or "Ecstasy"). Psychopharmacology. 1995;119:247–60.
116. Ashgar K, De Souza E. Pharmacology and Toxicology of Amphetamine and Related Designer Drugs. Rockville, MD: US Department of Health and Human Services, Public Health Service, Alcohol, Drug Abuse, and Mental Health Administration; 1989.

Part IV
Designer Drugs and their Abuse

Designer drugs are synthetic (man-made), illegal drugs produced in underground labs and sold on the street. Many of these drugs are used by youth and young adults at dance parties and clubs, and therefore are also known as *"club drugs"*. Although users may think these drugs are harmless, they can be dangerous. It is impossible to know exactly what chemicals were used to produce them, and they are often used in combination with other drugs or alcohol with unpredictable and dangerous results. (Fig. 1)

Fig. 1 "Designing" new compound by simple molecular modification in order to circumvent the illegal mother drug being under the Controlled Substances Act

History of Designer Drugs

The term was originally coined in the 1980s to refer to various heroin-like synthetic substances, mostly based on the fentanyl molecule (e.g. α-methylfentanyl; Fig. 2). The term gained widespread popularity when MDMA (ecstasy) experienced a popularity boom in the mid-1980s. In the United States, the Controlled Substances Act was amended by the Controlled Substance Analogue Enforcement of 1986, which attempted to ban designer drugs pre-emptively by making it illegal to manufacture, sell, or possess chemicals that were substantially similar in chemistry and pharmacology to Schedule I or Schedule II drugs.

> SCHEDULE I: Schedule I is a listing of those substances, which are controlled under US federal laws, are deemed to have a high potential for abuse, and for which there is no accepted medical use.

> SCHEDULE II: Schedule II is a listing of those substances, which are controlled under US federal laws, are deemed to have a high potential for abuse, and for which there is an accepted medical use.

> SCHEDULE III: Schedule III is a listing of those substances, which are controlled under US federal laws. They have a potential for abuse less than the drugs or other substances in schedules I and II, have a currently accepted medical use in treatment in the United States, and may lead to moderate or low physical dependence or high psychological dependence.

SHEDULE IV: Schedule IV is a listing of those substances, which have a low potential for abuse relative to the drugs or other substances in schedule III. They are currently accepted for medical use in treatment in the United States, and any abuse of these substances may lead to limited physical dependence or psychological dependence relative to the drugs or other substances in schedule III

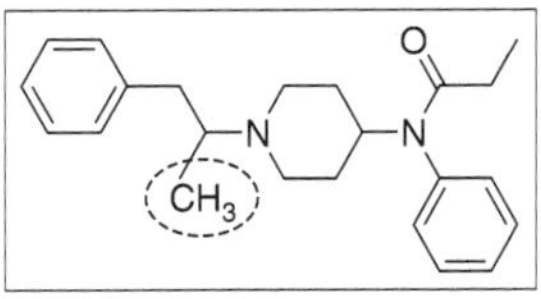

Fig. 2 By simple introduction of a methyl group, the illegal fentanyl resulted in the legal α-methylfentanyl

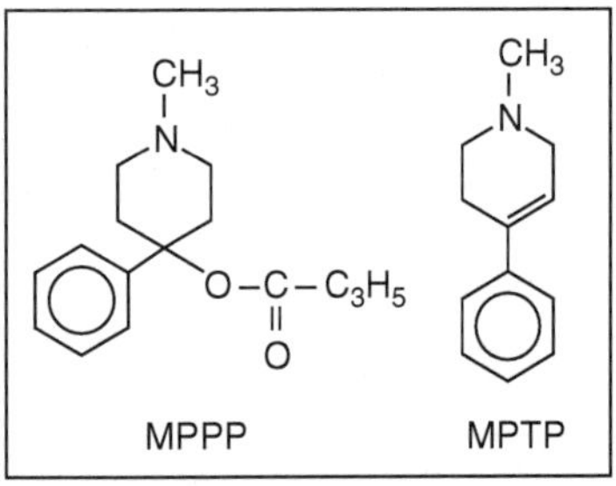

Fig. 3 Molecular structure of the meperidine analogue 1-phenyl-4-methyl-4-hydroxy-piperidine (MPPP) and its neurotoxic by-product 1-phenyl-4-phenyl-1–2–3–6-tetra-hydropyridine (MPTP)

Other countries have dealt with the issue differently. Germany, Canada, and the United Kingdom simply ban new drugs as they become a concern. Some countries, such as Australia and New Zealand, have gone the opposite direction and enacted sweeping bans based solely on the chemical structure, making chemicals illegal even before they are created. If a theoretical chemical fits a set of rules regarding substitutions and alterations of an already banned drug, it automatically is banned. The controlled substance analogue law in New South Wales, Australia, is so broad that it would cover millions of compounds that have never been made, simply on the basis that they bear a vague resemblance to one of the drugs on the illegal list. When the term was coined in the 1980s, a wide range of narcotics were being sold as heroin on the black market. Many were based on fentanyl or meperidine. One synthetic heroin or MPPP (Fig. 3), was found in some cases to contain an impurity called MPTP, which caused brain damage that could result in a syndrome identical to full-blown Parkinson's disease, from only a single dose [1, 2].

Other problems were highly potent fentanyl analogues, which were sold as "China White", that caused many accidental overdoses. Because the government was powerless to prosecute people for these drugs until after they had been marketed successfully, laws were passed to give the DEA power to emergency schedule chemicals for a year, with an optional 6-month extension, while gathering evidence to justify permanent scheduling, as well as the analogue laws mentioned previously.

Emergency-scheduling power was used for the first time for MDMA (ecstasy). In this case, the DEA scheduled MDMA as a Schedule I drug and retained this classification after review, even though their own judge ruled that MDMA should be classified Schedule III on the basis of its demonstrated uses in medicine. The emergency scheduling power has subsequently been used for a variety of other drugs including 2C-B, AMT, and BZP. In 2004, a piperazine drug, TFMPP, became the first drug that had been emergency-scheduled but was denied permanent scheduling and reverted to legal status.

In the late 1990s and early 2000s, there was a huge explosion in designer drugs being sold over the Internet. The term and concept of *"research chemicals"* was coined by some marketers of designer drugs, particularly of psychedelic drugs in the tryptamine and phenethylamine family. The idea was that by selling the chemicals as for research rather than human consumption, the intent clause of the US analogue drug laws would be avoided. This was later shown to be faulty logic when the DEA raided multiple suppliers, and multiple vendors several years later in a joint Web Operation. This process was accelerated greatly when vendors began advertising via search engines like Google by linking their sites to searches on key words such as chemical names and terms like psychedelic or hallucinogen. Widespread discussion of consumptive use and the sources for the chemicals in public forums also drew the attention of the media and authorities.

Many substances that were sold as *"research chemicals"* in this period of time are hallucinogens and bear a chemical resemblance to well-known drugs, such as psilocybin and mescaline. As with other hallucinogens, these substances are often taken for the purposes of facilitating spiritual processes (entheogen), mental reflection (psychedelic) or recreation. Some research chemicals on the market were not psychoactive, but can be used as precursors in the synthesis of other potentially psychoactive substances. For example 2C-H, which could be used to make 2C-B and 2C-I among others. Pharmaceutical corporations, universities and independent researchers over the last century, from which some of the presently available research chemicals derive, have conducted extensive surveys of structural variations. One particularly notable researcher is Dr. Alexander Shulgin (Fig. 4), who presented syntheses and pharmacological explorations of hundreds of substances in the books TIHKAL and PIHKAL (co-authored with Ann Shulgin), and has served as an expert witness for the defense in several court cases against manufacturers of psychoactive drugs [3, 4].

Most chemical suppliers sold research chemicals in bulk form as powder, not as pills, as selling in pill form would invalidate the claims that they were being sold for non-consumptive research. Active dosages vary widely from substance to substance, ranging from sub-microgram levels to hundreds of milligrams. While it is critical for the end user, instead of guessing to weigh doses with a precision scale, such precision scaling was abandoned. This led to many emergency room visits and several deaths, which were a prominent factor leading to the emergency scheduling of several substances and eventually Operation Web Type. When a chemical increases in popularity, it often will be sold in pill form to reach a wider market. Some of the most popular chemicals also are given street names like *"Foxy"* or *"Foxy Methoxy"*

Fig. 4 The chemist Alexander Shulgin who while working at Dow Chemical synthesized MDMA

for 5-Meo-DIPT. Once a chemical reaches this kind of popularity, it is usually just a matter of time before it is added to the list of scheduled (i.e. illegal) drugs.

Well-Known Designer Drugs of the Past Years

Most research chemicals are structural analogues of tryptamines or phenethylamines, but there are also completely unrelated chemicals. Thus, it is impossible to determine psychoactivity or other pharmaceutical properties of these chemicals strictly from examining their structure, and many of the substances have common effects while structurally different and vice versa. Confusing nomenclature, similar names, and differing naming schemes can all lead to and have led to potentially hazardous mixups for end users.

Some of the early narcotic designer drugs (Fig. 5) are:

- α-Methylfentanyl, which became well known as "*China White*" on the heroin market
- Parafluorfentanyl
- 3-Methylfentanyl

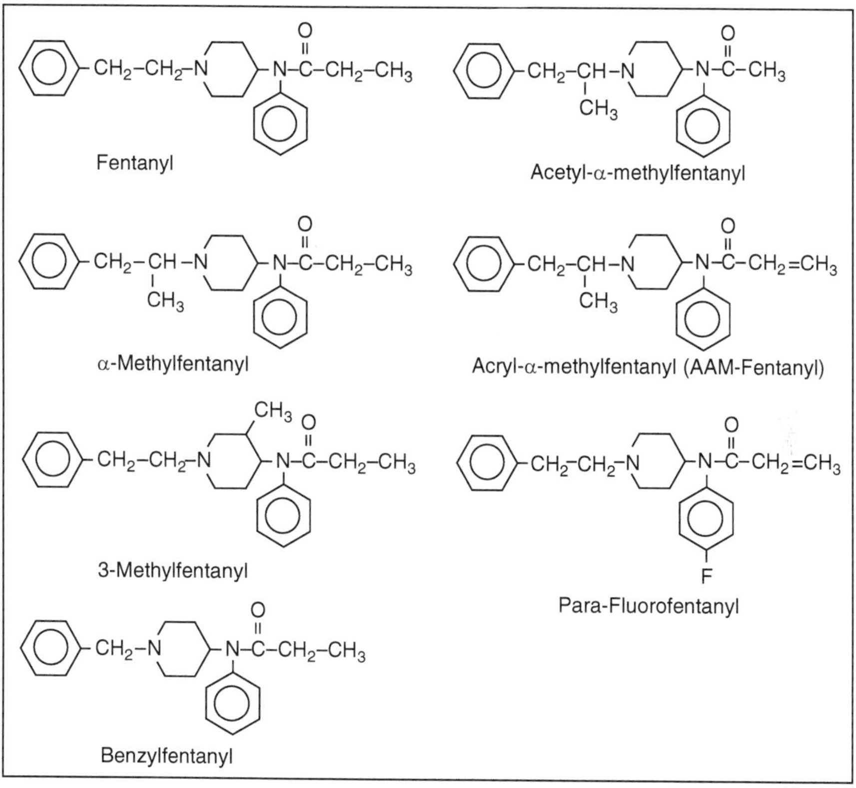

Fig. 5 Molecular structure of some fentanyl derivatives known as designer drugs, manufactured by clandestine laboratories

- MPPP became especially famous due to an impurity in some batches called MPTP (Fig. 3), which caused permanent Parkinsonism with a single use.

Especially the fentanyl analogues, because of their high purity and analgesic potency, caused respiratory depression following intake (Fig. 6).

On the other hand many other prominent tryptamine-based related designer drugs were synthesized:

- 4-Acetoxy-DIPT,N,N-diisopropyl-4-acetoxytryptamine
- 5-MeO-AMT, 5-methoxy-alpha-methyltryptamine
- 5-MeO-DIPT, 5-methoxy-diisopropyltryptamine, also known as "*Foxy*" or "*Foxy Methoxy*"
- 5-MeO-DMT, 5-methoxy-dimethyltryptamine
- AMT, α-methyltryptamine
- AET, α-ethyltryptamine
- DIPT, N,N-diisopropyltryptamine
- DPT, N,N-dipropyltryptamine

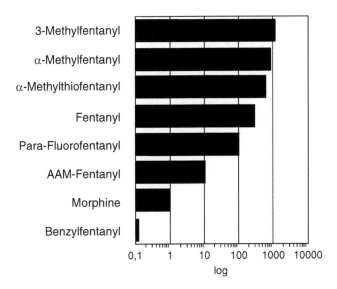

Fig. 6 Analgesic potency of various fentanyl analogues, which appeared on the streets and due to their activity to depress respiration resulted in several incidences of death

From the phenethylamine-based series the following prominent designer drugs appeared on the street:

- 2C-B, 4-bromo-2,5-dimethoxyphenethylamine
- 2C-C, 2,5-dimethyoxy-4-chlorophenethylamine
- 2C-I, 2,5-dimethoxy-4-iodophenethylamine
- 2C-E, 2,5-dimethoxy-4-ethyl-phenethylamine
- 2C-T-2, 2,5-dimethoxy-4-ethylthiophenethylamine
- 2C-T-7, 2,5-dimethoxy-4-(n)-propylthiophenethylamine
- 2C-T-21, 2,5-dimethoxy-4-(2-fluoroethylthio)phenethylamine
- MDMA, 3,4-methylenedioxymethamphetamine
- MDEA, 3,4-methylenedioxy-N-ethylamphetamine
- DOB, 2,5-dimethoxy-4-bromoamphetamine
- DOM, 2,5-dimethoxy-4-methylamphetamine
- TMA-2, 2,4,5-trimethoxyamphetamine

Also, PCP (phencyclidine) analogues were being sold as designer drugs, which were characterized of having a hallucinogenic effect:

- TCP, 1-[1-(2-thienyl)-cyclohexyl]-piperidine or thienyl-cyclohexyl-piperidine
- PCE, (1-phenylcyclohexyl)-ethylamine
- PCPγ, 1-(1-phenylcyclohexyl)-pyrrolidine

Only recently some newly designed piperazine-based drugs emerged:

- BZP, benzylpiperazine
- TFMPP, 3-trifluoromethylphenylpiperazine, has the unique distinction of being the only drug, which was emergency scheduled into Schedule I and then allowed to become legal because the DEA was unable to justify permanent scheduling
- mCPP, 1-(3-chlorophenyl)piperazine
- pFPP, 1-(4-fluorophenyl)piperazine

Rarely mentioned designer drugs are:

- The steroid tetrahydrogestrinone (THG, "The Clear"), recently was listed as an unapproved drug by the Food and Drug Administration (FDA), and cannot be legally marketed in the United States. The FDA is concerned about the use of this unapproved product. They warn consumers that little is known about the safety of this drug, and its use may pose considerable risks to health similar to other steroid hormones used by athletes. THG is considered a "designer" synthetic steroid derived by chemical modification from another anabolic steroid that is banned by the US Anti-Doping Agency. THG is structurally related to two other synthetic anabolic steroids, gestrinone and trenbolone.
- Madol, which sometimes confusingly is referred to as "DMT".
- GBL or γ-butyrolactone, is a precursor to and substitute for γ-hydroxybutyric (GHB), which is converted in the body to GHB (Fig. 7).
- 1,4-Butanediol (1,4-BD), another GHB analogue (Fig. 8) emerged, since γ-hydroxybutyric acid (GHB), a neuroprotective therapeutic drug became illegal in a number of countries. This agent is also converted in the body to GHB.

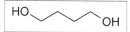

Fig. 8 Molecular formulary of 1,4-butanediol (1,4-BD), which is metabolized in the body to γ-hydroxybutyric acid (GHB)

Gamma-Hydroxybutyric (GHB) Acid

This new agent became known as "Liquid Ecstasy, Liquid E, Liquid X, or Fantasy". It is a central nervous system depressant, and has been used as an anxiolytic, anesthetic and sedative/hypnotic, inducing a sleep-like state in experimental animals in doses ranging from 0.1 to 1.5 g/kg [5]. GHB is also a putative neurotransmitter [6]. Treatment with GHB has been shown to affect many neurotransmitter systems, most notably, dopamine, glutamate and acetylcholine [7]. A number of early studies focused on GHB's action at dopaminergic synapses and showed that GHB initially inhibits dopamine release and then, time-dependently, causes its release [8]. Importantly, GHB exhibits an extremely high affinity for a specific GHB receptor [9]. This GHB receptor is a novel G-protein coupled receptor [10], which is widely distributed throughout the cortex and certain subcortical locations, most notably the hippocampus, amygdala, septum, basal ganglia and substantia nigra [11]. It is obvious that GHB receptors mediate some of GHB's action because cortical neurons, which are normally activated by GHB revert to their normal activity levels when GHB is added, along with its selective receptor antagonist NCS-3827 [11a]. The recent development of truly selective GHB receptor agonists has allowed scientists to finally understand GHB's mechanism. At low doses, GHB is binding almost exclusively to the GHB receptor, a receptor, which through pathways not yet understood, stimulates the release of glutamate, the major excitatory neurotransmitter in the cortex. As the dose of GHB increases, more and more GHB binds to, and activates the inhibitory $GABA_B$ receptor. This leads to the sedative/hypnotic effect of GHB. GHB's dose-response curve is similar to that of alcohol. At higher doses, users fall unconscious and are temporarily unable to be awakened (coma). It also may dangerously depress breathing, therefore mixing with alcohol should be avoided.

History of GHB

In the 1960s, a French researcher synthesized GHB in an attempt to create γ-aminobutyric acid (GABA) analog that would, unlike GABA, cross the blood-brain barrier [12]. Somewhat later, in 1963, GHB was found to be a naturally occurring metabolite in the human brain [13]. The first accepted medical application of GHB was for intravenous induction of anesthesia [12]. However, its use was limited due to the high frequency of vomiting [14], seizure-like activity in animals [15, 16] and inability to produce analgesia [17]. In the 1970s, GHB was recommended for narcolepsy because it increases slow-wave sleep and consolidates sleep at night, therefore decreasing sleep during the day [18]. In the 1980s, GHB was commonly sold over-the-counter in health food stores where it was alleged to increase the effect of growth hormone [19], while in the late 1980s and early 1990s, GHB was advocated for the treatment of alcohol dependence [20] and to counteract opiate withdrawal [21]. During the same time period, GHB was illicitly advertised as a hypnotic to replace tryptophan, which had been removed from the market due to its connection with eosinophilia-myalgia syndrome. Since 1990, an increasing number of cases of both abuse and toxic reaction has been noted, and in 1997 GHB was labeled a "date rape" drug by the press. In March 2000, GHB became a schedule I controlled substance in the US [22]. Nevertheless, γ-hydroxybutyric acid has become increasingly popular as a drug of abuse over the last 10 years. Many names are used for GHB such as sodium oxybate, sodium oxybutyrate, γ-hydroxybutyrate sodium, γ-OH, 4-hydroxy butyrate, and γ-hydrate, as well as others. Names used on the street include "Liquid Ecstasy, Liquid X, Liquid E, Georgia Home Boy, Grievous Bodily Harm, G-Riffick, Soap, Scoop, Salty Water, Somatomax, and Organic Quaalude".

Most of the GHB available in the US and Europe is manufactured clandestinely. Many Internet sites and books that describe the process of making GHB are available [23]. Commonly offered for sale on Internet sites, GHB kits provide the chemicals and recipes used to produce GHB [24, 25]. Currently, GHB is only legally available in the US for the investigational treatment of narcolepsy. The drug is synthesized by using a combination of sodium hydroxide and γ-butyrolactone or GBL, another commonly abused drug (see page 265). Because sodium hydroxide is very caustic, severe toxic reactions may result if GHB is manufactured improperly. The drug GHB is available as a powder or a colorless, odorless liquid with a salty or soapy taste. Adding it to flavored beverages can easily mask its taste. As GHB is colorless and odorless, and because small quantities are required to achieve a desired effect, GHB has been used as a "*date rape drug*" (Fig. 9). The amnesia produced by GHB often makes victims unable to serve as valid witnesses. Due to the frequency of GHB abuse, the drug was banned in the 1990s in both the United States and Canada.

Medical Use of γ-Hydroxybutyrate

In the 1960s, a synthetic form of GHB was created in order to treat sleep disorders and other health conditions, and GHB has become available in the United States under the brand name Xyrem®. In 2002 the FDA allowed Xyrem® (or sodium oxybate 4.5 g from Orphan Medical, a subsidiary of Jazz Pharmaceuticals), to be used in order to treat severe sleep disorders including narcolepsy and cataplexy. Physicians also have begun to prescribe Xyrem® for other purposes, including:

Fig. 9 Liquid Ecstasy or γ-hydroxybutyric acid being used repetitively as a "date rape" drug in social gatherings

- Insomnia
- Cataplexy associated with narcolepsy
- Fibromyalgia
- Alcoholism
- Addiction

Beginning in the late 1990s, researchers found a surprising link between the use of Xyrem® and a dramatic decrease in fibromyalgia symptoms of muscle pain and fatigue. It is now believed that GHB can significantly reduce fibromyalgia pain and fatigue by increasing the amount of slow-wave sleep that fibromyalgia patients experience on a nightly basis because between 50% and 100% of fibromyalgia patients suffer from a sleep disorder known as alpha EEG anomaly. This sleep disorder prevents sufferers from experiencing the normal amount of slow wave sleep (also known as deep sleep). As a result, fibromyalgia patients suffer from extreme fatigue. Additionally, it is during slow-wave sleep that the body works to release toxins from the bloodstream and heal damaged muscle and nerve cells. Because fibromyalgia patients do not get enough slow-wave sleep, their bodies do not have the chance to heal properly, exacerbating other symptoms. It is believed the GHB can help to increase slow-wave sleep, thereby improving fatigue and other fibromyalgia symptoms. To date, a number of different studies have been performed using GHB as a fibromyalgia treatment. The first study, which took place in 1998, involved 11 fibromyalgia patients. These patients were administered Xyrem® over a period of 4 weeks and then evaluated for fibromyalgia symptoms. At the conclusion of the study, all patients reported decreased pain and fatigue.

Presently a phase III trial is under way evaluating the efficacy of GHB in lifestyle, physical activity, the reduction of pain and fatigue in adults with fibromyalgia. This was followed by a joint license agreement between UCB (Union Chemie Belge) and Jazz Pharmaceuticals, Inc. for an expanded marketing of Xyrem® (sodium oxybate). Under the agreement, UCB obtains the right to commercialize Xyrem®

for the treatment of fibromyalgia syndrome in countries outside the US (where it is a Schedule III drug), when the product is approved for this indication.

Pharmacokinetics of GHB

The pharmacokinetics of GHB are nonlinear in humans over the therapeutic dosage range [26–29]. The drug is rapidly absorbed orally with an onset of action within 15 min [30, 31]. While in the rat, oral bioavailability is 52%–65% [32], in humans, the free fraction of GHB in plasma has been shown to be 0.99, indicating a lack of significant plasma protein binding [26]. The half-life of GHB is 22 min–28 min after an oral dose of GHB 25 mg/kg [26, 27], with a half-life that is slightly longer with higher doses. In one study, GHB exhibited a longer half-life of 53 min in patients with narcolepsy after dosages of GHB 3.0 g twice/night, administered 4 h apart [28]. At lower doses of 25 mg/kg, the Tmax of GHB is approximately 30 min, and after higher doses of 50 mg/kg, the Tmax occurs around 45 min [26, 27]. As the dose of GHB increases by a factor of four from 12.5 mg/kg to 50 mg/kg, the area under the cure (AUC) increases by a factor approaching seven. In addition, the Cmax increases, but not to the degree expected in relation to the increase in AUC and the decrease in oral clearance. The oral clearance is halved from 14 ml/min/kg to 7 ml/min/kg when the dose is increased from 12.5 mg/kg to 50 mg/kg, respectively.

Two suggested mechanisms for this nonlinearity include the saturation of one of the metabolic pathways of GHB or the capacity-limited absorption of GHB [26]. The capacity-limited absorption would explain the relatively small increase in Cmax with increasing dose, relative to the decrease in oral clearance. It is possible that several mechanisms operate concurrently, explaining the pharmacokinetic profile of GHB. In one investigation, four of the five subjects exhibiting linear kinetics had normal liver function test results, whereas all of the subjects who displayed nonlinear kinetics (capacity-limited elimination) had elevated values for some of their liver function tests [27]. In the subjects displaying nonlinear kinetics, a doubling of the dose resulted in a disproportionate threefold increase in the AUC from approximately 3,500 to approximately 10,800 µg/ml/min. The appearance of nonlinearity was found only in the patients who had abnormal liver function values. The authors suggest a relationship between liver function and saturation of the elimination pathway for GHB. In addition, when the dose was increased from 25 mg/kg to 50 mg/kg, there were proportional changes in Cmax, accompanied by a disproportionate increase in AUC.

The Pharmacology of γ-Hydroxybutyric Acid (GHB)

The exact mechanism of GHB action in the CNS has not been determined, but GHB is structurally related to GABA (Fig. 10), which is a precursor in GHB formation [33]. Much debate exists regarding whether GHB has neurotransmitter or neuromodulatory roles [34], because GHB has high-affinity brain receptors and undergoes synthesis, release, uptake, and degradation within the CNS [35–37]. The exact location of the biosynthetic pathway of GHB inside the cell (cytosol vs mitochondria) has not been fully established [38], as the neurotransmitter GABA is transaminated by GABA amino-transferase to form SSA (succinic semialdehyde), which is either further metabolized into succinic acid or reduced to form GHB by the enzyme SSA reductase, a NADPH-dependent enzyme [39, 40]. Also, the highest concentrations of GHB in the brain are found in the substantia nigra [41] and the hypothalamus, whereas the highest turnover rate of GHB occurs in the hippocampus [42]. In addition, the uptake of GHB appears to be the highest in the striatum [43, 44], which is dependent on a specific sodium-dependent active transport system for GHB [43]. Such an active transport system for GHB is also being found in the kidney, heart, skeletal muscle, and brown fat [45] although its significance is not clear yet.

Mode of Action of GHB

γ-Hydroxybutyrate appears to have affinity for two receptor sites in the CNS. It binds to GHB receptors, which may be linked to cyclic guanosine 3,5-monophosphate and inositol phosphate intracellular pathways [46, 47] and are most numerous in the hippocampus and cortex [11]. There it binds to $GABA_B$ [48, 49] but not to $GABA_A$ receptors [50]. Although the relevance of this remains unknown it suggests that some of the pharmacologic actions of GHB are mediated by the $GABA_B$ receptor (Fig. 11).

The drug GHB alters dopaminergic activity, in some cases increasing and in others, decreasing the amount of dopamine released [38]. The systemic administration of GHB to animals results in increased dopamine accumulation in the extrapyramidal system of the brain, which reaches its highest values 1–2 h after injection, without

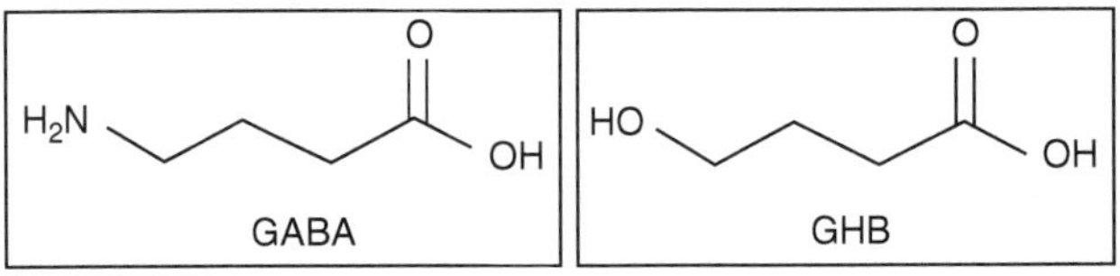

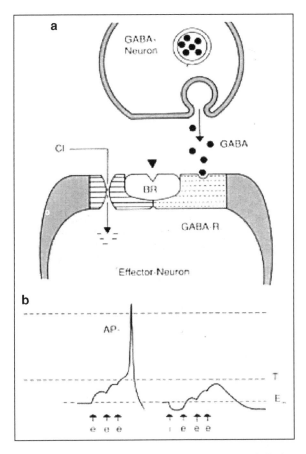

Fig. 10 Molecular formula of the natural neurotransmitter gamma-amino-butyric acid (GABA) and gamma-hydroxy-butyric acid (GHB)

Fig. 11 Potential mode of action of GHB and GBL, both of which mimic the action of GABA (γ-aminobutyric acid; 11a), an inhibitory neurotransmitter at the CNS resulting in a depression of action potentials (AP; 11b)

parallel increases in serotonin or norepinephrine [51]. The administration of α-methyltyrosine, which blocks the activity of tyrosine hydroxylase, almost completely blocks the rise in brain dopamine induced by GHB, which occurred within 1 h in control mice. Therefore it can be concluded that GHB mediates an accumulation of dopamine by increasing the activity of tyrosine hydroxylase [7]. In addition, GHB may inhibit the release of newly synthesized dopamine [52] and decrease the firing rate of dopaminergic neurons in the substantia nigra with maximal inhibition within 8 min [41].

The net result seems to be a tissue accumulation of dopamine in the brain [36], which is supported by results of the short-term studies described above. Dopamine release in the striatum may be accompanied by the release of endogenous opioids [53]. The exact interactions between GHB and the opioid system are not fully understood, but the administration of naloxone or nalorphine, opioid receptor antagonists, blocks some of the effects of GHB [53, 54].

Dose-Related CNS Effects of GHB

The primary dose-related effects of GHB are related to CNS depression. At 10 mg/kg, GHB is capable of producing amnesia [55] and hypotonia of the skeletal muscles [56] resulting from the depression of neurons in the spinal cord [31]. At 20–30 mg/kg, GHB promotes a normal sequence of rapid eye movement (REM) and non-REM (slow-wave) sleep, which lasts from 2 to 3 h [18, 57]. At 40–50 mg/kg intravenously, GHB produces a state of somnolence, which appears within 5–15 min, and an oral dose of approximately the same amount will produce similar results [31]. Anesthesia is associated with doses of 50 mg/kg [29, 31, 58] and doses higher than 50 mg/kg have been associated with profound coma [58] as well as decreased cardiac output, respiratory depression, and seizures. These effects are more pronounced with the co-ingestion of CNS depressants, particularly ethanol [59]. Larger doses of 60–70 mg/kg produce a state of unarousable coma that lasts about 1–2 h [31]. The investigators, who initially discovered that GHB was a natural metabolite of the brain reported that GHB 100 mg/kg administered intravenously produced sleep that begins within 15 min of administration and lasts about 1.5–2 h [13].

Oral ingestion of GHB 75–100 mg/kg in humans results in peak blood levels of approximately 90–100 μg/ml at 1–2 h after ingestion. Intravenous administration of GHB 50 mg/kg and 165 mg/kg results in peak blood levels that reach 180 μg/ml and 412 μg/ml, respectively. The mean blood GHB level at the commonly used dose of 100 mg/kg is 304 μg/ml [29]. When the blood GHB levels exceed 258 μg/ml, subjects fall into a state of deep sleep, characterized by nonresponse to various stimuli. However, there is still a reflex response to surgical incision. During this stage of deep sleep, blinking stops and the eyes remain central and fixed with small pupils. A moderate level of sleep is associated with blood GHB levels ranging from 155–258 μg/ml. Spontaneous blinking and responses to deep pressure characterize this moderate stage of sleep. Blood GHB levels ranging from 52 μg/ml to 155 μg/ml are associated with a light sleep accompanied by spontaneous movements and occasional opening of the eyes. When the blood GHB levels decrease below 52 μg/ml, subjects wake up [29].

Cardiovascular Effects of GHB

Moderate bradycardia appears after the administration of GHB [17, 103] and is likely due to central vagal activity [31]. In addition to bradycardia, GHB reduces stroke volume as well as cardiac output, which reaches a nadir around 30 min after ingestion. Atropine reverses the decreases in both heart rate and stroke volume [60]. The autonomic centers are fully active during GHB-induced coma, and surgical stimuli result in a cardiovascular response, such as tachycardia, hypertension, and raised cardiac output [17, 31, 61].

Respiratory Effects of GHB

Respiratory rate is often reduced, but this is usually accompanied by an increase in tidal volume [17,61]. The drug GHB also produces a slowing and deepening of respiration sometimes leading to a Cheyne-Stokes pattern [31,61].

Neuroendocrine Effects of GHB

In an early study that stimulated much interest in the use of GHB by the body-building population, intravenous administration of GHB 2.5 g significantly increased plasma growth hormone levels, which peaked at 60 min [19]. In a more recent study, after bedtime oral ingestion of GHB 2.5, 3.0, and 3.5 g, a significant increase occurred in the normal secretory pulse of growth hormone during the first 2 h after sleep onset. The authors suggest that agents such as GHB may increase the release of growth hormone by increasing slow-wave sleep, because there is a large pulse in growth hormone secretion during the first stage of slow-wave sleep more than 90% of the time [62].

Sleep Pattern Following GHB

The drug GHB stimulates slow-wave sleep [58, 63, 64]. It does not appear to suppress REM sleep [63, 65] and may even decrease fragmentation of REM sleep [66]. It appears to increase "slow" sleep as evidenced by a slow synchronized electroencephalographic recording [58]. In addition, GHB increases slow-wave sleep (stages 3 and 4), whereas light sleep (stage 1) is decreased, and the frequency of awakenings is reduced [64]. In healthy subjects, under double-blind conditions, single oral doses of GHB 2.25 g significantly increased the time spent in slow-wave sleep, while sacrificing stage 1 sleep and significantly decreasing slow-wave sleep latency. The efficiency of REM sleep is increased, but the REM latency and time spent in REM sleep do not change [67].

Use of GHB in Clinical Medicine

GHB for Sedation and Anesthesia

Most of the therapeutic applications of GHB result from its sedative and hypnotic effects on the CNS. There are no currently accepted medical applications for GHB in the US, although it is being evaluated for the symptoms of narcolepsy. However, GHB has been extensively administered and studied for a variety of indications in other countries in Europe. Although the principal actions of GHB have not been fully elucidated, results of early investigations suggest that GHB appears to act on the cerebral cortex with little or no depression of the reticular activating system [17]. Some authors speculate that there is depression of the limbic hippocampal structures [31] and subcortical centers [58]. The anesthetic effects of GHB are primarily hypnotic [61] as GHB provides little or no analgesia [17, 61]. The transition from wakefulness is described as being a sudden shift from responsivity to unconsciousness [58].

The first clinical application of GHB was as a hypnotic anesthetic agent [12]. It is still given for sedation and anesthesia in Germany, where it is considered safe and effective as long as the doses given are limited to the clinical needs [68]. In doses of 10–20 mg/kg, GHB demonstrates hemodynamic stability and lack of severe respiratory depression, while control and recovery are acceptable for clinical purposes [69]. However, bradycardia, hypotension, arrhythmias, and severe respiratory depression have been reported during GHB intoxication (see section on Adverse Effects).

GHB for Cerebral Protection

γ-Hydroxybutyrate may be an endogenous inhibitor of energy metabolism, protecting tissues when energy supplies are low. Evidence suggests that GHB reduces cellular activity, while depressing the utilization of glucose as well as other energy substrates.

This may result in tissues being less sensitive to the damaging effects of anoxia or during periods of excessive metabolic demand. Therefore, the natural function of GHB may include a role as a tissue protective substance [70]. γ-Hydroxybutyrate reduces tissue oxygenation demands and protects cells during hypoxic states, which has been demonstrated in both human and animal studies as well as in various organ systems. It exerts a protective effect and reduces cellular damage during sepsis, hemorrhagic shock, great vessel or coronary artery occlusion, stroke, organ transplantation, and myocardial infarction. In addition, in humans with brain tumors, GHB decreases intracranial pressure and increases cerebral blood flow. A thorough review of these topics involving the cellular protective effects and cerebral protective effects of GHB, as well as various applications for GHB in anesthesia, has been published [71].

GHB in Narcolepsy and Insomnia

Owing to the ability of GHB to increase slow-wave sleep and facilitate REM sleep efficiency, GHB may improve nighttime sleep and therefore improve alertness during the day, which could alleviate some of the symptoms of narcolepsy [18, 63–68]. In addition, administration of GHB to patients with narcolepsy revealed significant improvements in sleep attacks, daytime drowsiness, cataplexy, hypnogogic hallucinations, and sleep paralysis [65, 72]. Because GHB is a CNS depressant, it has been investigated for treating the symptoms of insomnia [30] and in one investigation it was rated by the subjects as being an "excellent hypnotic" [18]. However, when being used as a hypnotic, an oral dose of GHB 100 mg/kg resulted in frequent awakenings at either 1.5 or 4–5 h after ingestion, which accounted for 14 of the 25 adverse effects reported in this dose group [30]. Furthermore, GHB reportedly produced sleep paralysis, sleep walking, and cataplexy [30].

GHB in Alcohol and Opiate Withdrawal

The drug GHB 50 mg/kg/day has been given orally to treat the symptoms of acute alcohol withdrawal and to facilitate both short- and long-term abstinence from alcohol. It also was given to treat opiate withdrawal, often in higher dosages of 50–300 mg/kg/day. These applications of GHB were discussed extensively in a review of this topic during a symposium hosted by the Italian Society on Biological Psychiatry [73]. Despite a possible benefit of taking GHB for these conditions, craving for GHB developed during these trials, with some subjects increasing their dosage up to 6–7 times the recommended levels [74].

Metabolism of γ-Hydroxybutyrate (GHB)

The biosynthetic pathway and metabolic degradation of GHB occurs in brain tissue by means of multiple cytosolic and mitochondrial enzymes. Gamma-hydroxybutyrate is a natural product of GABA metabolism by way of the intermediate compound, succinic semialdehyde (SSA). The neurotransmitter GABA appears to be the major precursor for SSA, from which GHB is synthesized. In early investigations, it was shown that GABA undergoes transamination to succinic semialdehyde or SSA [75] that the enzymatic reduction of SSA to GHB occurred in mammalian brain in vitro [76], and that [³H]GABA was converted to GHB in the rat brain in vivo [76]. Other investigators [39] confirmed the finding that GABA is a precursor of GHB. The NADPH-dependent enzyme SSA reductase is responsible for the conversion of SSA to GHB [39] (Fig. 12). Gamma-hydroxybutyrate is oxidized to SSA by means of GHB dehydrogenase and GHB-oxoacid transhydrogenase [77], which is further metabolized to succinate, then entering the Krebs cycle [40].

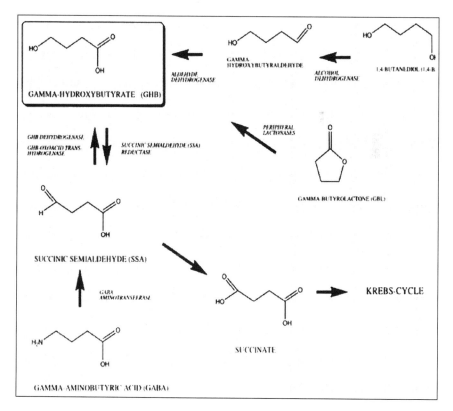

Fig. 12 Mechanism of GHB elimination and the conversion of the precursors γ-butyrolactone (GBL) and 1,4-butanediol (1,4-BD) in the body to GHB

Less than 2% of GHB is eliminated unchanged in the urine [27, 29]. Owing to the short half-life, there is no accumulation of GHB with repeated dosing and GHB doses of up to 100 mg/kg are no longer detectable in the blood after 2–8 h or in the urine after 8–12 h [27, 30]. The variability of these findings may depend on the sensitivity of the assay used, or it may be due to interindividual variability. In summary, it has been suggested that regardless of the dose given, the elimination of GHB is so rapid, even in those with compromised liver function, that the drug is completely eliminated within 4–6 h after ingestion [27].

Abuse Potential and Intoxication with GHB

Factors that seem to contribute to the abuse potential of GHB include its intoxicating effects, its purported anabolic effects, its hypnotic effects, and its ability to incapacitate women for purposes of sexual assault [78, 79]. One of the main reasons GHB became a popular drug of abuse is its ability to produce a "high" [59, 80]. Those who take GHB describe it as producing a state of relaxation and tranquility accompanied by feelings of calmness, mild euphoria, a tendency to verbalize, mild numbing, and pleasant disinhibition. Despite these positive feelings attributed to the use of GHB, the dose-response curve for GHB has been described as being remarkably steep. Therefore, as the dose of GHB is increased, a steep increase in adverse effects may occur [81]. The effects of GHB have been described as being similar to those of alcohol, and the two agents may act synergistically, further increasing the risk for intoxication or overdose [82].

Although GHB is commonly taken for its proposed anabolic effects (related to the ability of GHB to stimulate the release of growth hormone), especially by the body-building community [19, 62, 83] no definitive evidence exists that it increases muscle mass or fat catabolism. In addition, in patients with chronic alcoholism, long-term administration of GHB did not affect muscular mass [84].

Incidence of Adverse Effects in GHB Intoxication

The Centers for Disease Control and Prevention (CDC) released two reports describing the toxic effects of GHB [85, 86]. These reports document over 120 poisonings and one fatality in individuals from various regions of the US who became ill secondary to taking GHB. The usual course of illness was very similar from case to case. Approximately 15–60 min after ingestion, one or more of the following symptoms occurred: vomiting, drowsiness, soporific state, hypotonia, or vertigo. Depending on the dosage taken and concurrent use of other CNS depressants, such as alcohol, any of the following occurred as well: loss of consciousness, respiratory depression, tremors, myoclonus, seizure-like activity, bradycardia, hypotension, or respiratory arrest. In many of these cases, the symptoms spontaneously resolved within 2–96 h [85, 86]. As a result of the increased rate of GHB abuse since the first CDC report in 1990, the number of acute intoxications due to GHB has increased [23, 25, 78, 80, 86–96]. Some of the more common and better documented conditions that appear in various reports include coma, respiratory depression, seizure-like activity (uncontrollable or unusual movements), bradycardia, drowsiness or dizziness, confusion, amnesia, headache, nausea, vomiting, mild hypothermia, acidosis, and psychiatric complications (e.g., agitation, delirium).

Since 1992, the DEA has documented over 9,600 adverse reactions, overdoses, and other cases reported by various law enforcement agencies, poison control centers, and hospitals in 46 states. The Food and Drug Administration has issued warnings to inform consumers about the dangers of ingesting two potentially dangerous GHB precursors, γ-butyrolactone (GBL) and 1,4-butanediol (BD), which are converted to GHB in the body [97]. The doses of GHB that elicit adverse effects vary greatly from report to report and range from a 0.25 teaspoon (1.25 ml) to a 4 tablespoons (60 ml) [78] up to 16 oz (480 ml) [87]. However, GHB often is produced in clandestine laboratories, resulting in preparations with a wide range of purity. Therefore, the ingested quantities reported in cases of acute intoxications may not be that informative. A 99% pure sample of GHB weighs 2.8 g/level teaspoon (5 ml) [87]. However, 40 ml of clandestinely produced GHB may weigh from 3–20 g [88]. One aspect of GHB that makes its use dangerous is that the response to oral ingestion seems to vary within the same person as well as between persons.

Principles in GHB Intoxication

The following adverse effects were found in experimental and clinical investigations, and in reports of intoxications. The drug GHB primarily affects the CNS, the cardiovascular, and the respiratory system. It however, has no toxic effect on the kidneys or the liver [18, 31]. Drowsiness and dizziness induced by GHB are reported frequently in both investigational and toxicity reports. Subjects receiving oral doses of GHB 25–50 mg/kg in a controlled study complained of dizziness and drowsiness [26]. Other common CNS adverse effects include vertigo and headache [30]. More serious CNS depression during intoxication with GHB commonly occurs. Numerous reports of intoxication with GHB describe patients who present with Glasgow Coma Scale (GCS) scores as low as 3–5 [88, 89–93, 98]. Recovery appears to be inversely related to GCS score, with a lower GCS score resulting in a longer time to recover [92]. Coma induced by GHB usually appears rapidly after ingestion, followed by a rapid and apparent full recovery. Often in the cases of intoxication, the unconsciousness will resolve within 6–7 h [87, 89–93, 98, 99]. One of the distinctively characteristic aspects of GHB intoxication is the rapid recovery, which is often uneventful and may create a false sense of security in the user [87].

Effects of Acute GHB Intoxication

Fatalities have been associated with GHB use [86, 100–102]. The DEA has collected investigative, toxicology, and autopsy reports from cases in which GHB was found in biological samples of the deceased. Since 1990, the DEA reports that they are aware of 68 deaths associated with the use of GHB, most of which have occurred in recent years. Details of the cases are not given. In an article discussing pre- and postmortem GHB blood and urine levels, the authors refer to four fatalities attributed to the use of GHB [101]. Three of the fatalities had postmortem blood GHB levels ranging from 52–121 mg/L. In a series of forensic samples submitted for laboratory analysis, blood GHB levels ranging from 3.2–168 mg/L were found in 15 of 20 autopsy specimens, although the deaths were not thought to be GHB related. Furthermore, GHB was not found in samples from living subjects who did not take GHB. Because of these findings, the authors suggest that GHB may be a natural product of postmortem decomposition occurring in blood [101]. Other investigators suggest that the magnitude of GHB levels found in many fatality cases is too significant to be attributed to postmortem decomposition [102].

Cardiovascular Effects with GHB Intoxication

Bradycardia has occurred when GHB was given for anesthesia [17, 31, 61] as well as in overdose situations [88, 92, 103]. In a retrospective review on GHB intoxication, 36% of patients had pulse rates defined as bradycardia (heart rate < 55 beats/min) and one patient required a single dose of atropine for a heart rate of 24 beats/min [92]. In the same case series, ten patients had hypotension (systolic blood pressure 90 mmHg) at presentation. Six of the patients with hypotension also had concurrent bradycardia, and in all six cases alcohol and/or another drug of abuse were present. In another case series of seven patients, the authors reported that five patients developed U-waves on their electrocardiograms after GHB exposure, although none of them was significantly hypokalemic. Three of these five patients had significant abnormalities that included first-degree heart block, right bundle branch block, and ventricular ectopy [95].

Respiratory Effects with GHB Intoxication

Respiratory depression, difficulty breathing, and apnea have been reported after the administration of GHB [31, 71, 78, 80, 86, 88, 92, 96, 103]. The respiratory depression may be very severe, and in some cases the respiratory rate may drop to as low as 4 breaths/min [88, 103]. Abnormal patterns of breathing such as Cheyne-Stokes breathing may result [61].

Psychotic Effects with GHB Intoxication

Under the influence of GHB, some individuals may become hostile, belligerent, and agitated [87, 89]. Patients display loss of consciousness and are extremely combative when stimulated, despite profound respiratory depression. Furthermore, they may require physical restraints to protect themselves and hospital personnel [95]. Psychiatric complications such as delirium, paranoia, depression, and hallucinations have been reported in a small number of patients [23, 104].

Ocular Effects with GHB Intoxication

During intoxication with GHB, the pupils have been described as being miotic and sluggishly reactive to light, and during coma induced by GHB the eyes have been found to be miotic and unresponsive to light [58, 89].

Metabolic Effects with GHB Intoxication

Mild acute respiratory acidosis is a common finding when GHB has been used as an anesthetic, as well as when it has been abused [92, 95]. In one review, 93% of patients had a pH less than 7.40, and 30% had a pH less than 7.30. In addition, 70% of patients had a partial pressure of carbon dioxide of 45 mmHg or greater [92].

Gastrointestinal Effects with GHB Intoxication

A high frequency of vomiting is associated with the use of GHB [14, 105] especially during induction and on emergence from intravenously induced anesthesia [31, 105]. In an early investigation, 52% of patients receiving GHB for anesthesia experienced nausea or vomiting [61]. According to one case series, vomiting was also very common and occurred in 30% of 88 cases of GHB intoxication. It typically occurred as the patients were regaining consciousness [92]. In another review of 78 cases of GHB overdose, vomiting was reported in 22% of the cases [98].

Body Temperature in GHB Intoxication

Although hypothermia has not been a universal finding during GHB intoxication, mild hypothermia has been observed in patients after a GHB overdose [89, 91–95]. In one study of 70 patients, 31% had an initial body temperature of less than 35°C, and the mean body temperature was 35.8 ± 1.1°C [92]. In an additional small series of five patients, hypothermia was reported in three patients, with the lowest temperature being 32.8°C [91].

Muscular Disorders in GHB Intoxication

There have been many reports of unusual, random clonic movements and uncontrollable shaking associated with GHB use [17, 31, 78, 80, 89, 91, 96, 98, 104]. In anesthesia studies, abnormal movements occurred during induction with GHB but were not accompanied by any seizure-like electroencephalographic tracings and could be reduced by administering a phenothiazine drug [17]. Administration of GHB will not necessarily result in abnormal epileptiform electroencephalographic changes [68] or seizure-like activity [92].

Miscellaneous Effects in GHB Intoxication

Cold and feeling of heavy extremities have been reported after oral ingestion of GHB 50 mg/kg [30]. Diaphoresis was reported in 35% of the 78 cases of GHB overdose in one investigation [98]. Home brewing of GHB, often from kits sold on Internet sites or from mail order sources, can lead to various adverse effects due to improper manufacturing of GHB. The manufacture of GHB involves the mixture of γ-butyrolactone and the alkaline substance, sodium hydroxide. The inappropriate manufacture of GHB may lead to a very alkaline mixture, resulting in esophageal damage [106]. In New York, a 20-year-old man aspirated during vomiting, resulting in damage to his lung tissue that was attributed to the mixture of gastric contents containing sodium hydroxide [86]. Hematuria has also occurred after the ingestion of improperly manufactured GHB. Here, home-brewed GHB was being made with swimming pool chlorine tablets instead of the required sodium hydroxide [24].

Withdrawal and Tolerance Following Use of GHB

Data gathered by the DEA indicate that those who take GHB have exhibited chronic self-administration, compulsive abuse regardless of adverse consequences, as well as drug-seeking behaviors. These data suggest individuals may become psychologically dependent on GHB. Physical dependence may develop, with a withdrawal syndrome occurring on abrupt discontinuation [79, 107–109]. Tolerance to the effects of GHB results in an increase in dosage and a withdrawal syndrome on cessation of GHB ingestion [79, 108–110]. The withdrawal syndrome is characterized by insomnia, tremor, and anxiety that may last approximately 1 week [79]. In addition, more severe symptoms have been reported, including confusion, hallucinations, delirium, and autonomic stimulation with tachycardia. The symptoms of withdrawal may begin within 1–6 h after the last dose of GHB and may last from 5 to 15 days [108].

One case of Wernicke-Korsakoff syndrome has been attributed to the use of GHB [111]. According to the authors, the patient had not imbibed alcohol for several months before admission, although there was no mention of an ethanol screen. The patient presented with the classic triad of symptoms of Wernicke-Korsakoff syndrome: global confusion, sixth nerve palsies, and ataxic gait. In addition, paranoid delusions and hallucinations were present. According to the authors, the atypical mental features represented GHB withdrawal and were similar, in part, to delirium tremens without the serious autonomic dysfunction. The patient's symptoms resolved quickly with thiamine treatment, with the eye movement abnormalities resolving rapidly, followed by resolution of the abnormal gait and mentation [111]. The clinical picture of GHB withdrawal appears to range from anxiety, tremor, and insomnia to more severe symptoms such as disorientation, paranoia, hallucinations, tachycardia, and possibly extraocular motor impairment.

Treatment Following GHB Intoxication

The mainstay of treatment for GHB intoxication is protection of the airway and assisted ventilation if needed. Intubation, to protect the airway, is a common treatment procedure during GHB intoxication, and assisted ventilation may be required in some cases [112, 113].

Laboratory monitoring should include serum electrolytes and blood glucose levels in symptomatic patients, and additional monitoring, such as pulse oximetry and arterial blood gases, in patients with respiratory depression. Because of the increased prevalence of abuse, GHB should be considered as a causal agent in any patient with coma of unknown origin at presentation. Since GHB is rapidly cleared from the body, it is often difficult to confirm the definite use of GHB. Because GHB will be missed by many conventional first-line urine drug screens [114], analysis with gas chromatography mass spectrometry is required for detection and quantification [115]. Therefore, a history from the patient, or others who witnessed the GHB use, may be important diagnostic information. However, because GHB has amnestic properties, the patient may not be able to provide a very reliable history. Some suggest that a history of body-building or athletic physique may aid in the diagnosis of GHB abuse as this drug is commonly used in this patient population [23, 90].

First-Line Treatment in GHB Overdose

The roles of gastric lavage and activated charcoal have been questioned as the volumes of GHB are very small and GHB is rapidly absorbed from the gut [116]. But these treatments may be helpful when GHB is coingested with other drugs of abuse.

Activated charcoal may be of benefit for recent, large ingestions of GHB [117]. Induction of emesis is not recommended because the CNS depression and diminished gag reflex may lead to pulmonary aspiration [117].

Because many of the symptoms of GHB intoxication are so rapidly reversed, it is difficult to determine whether purported helpful pharmacologic treatments have been successful or whether the GHB intoxication has simply worn off. In clinical cases of GHB intoxication, both naloxone and flumazenil have been found to be of no benefit in reversing unconsciousness [78, 85, 87–89, 93, 103, 104, 112].

Owing to the association between GHB and absence epilepsy in animals, various anticonvulsant agents have been used as GHB-reversal agents, but there are no data in the literature indicating that any of these agents have been useful in experimental or clinical situations in humans. Monitoring neurologic function and applying Glasgow Coma Scale (GCS) are essential. Patients with an initial GCS of 8 or less may have a more serious clinical course, requiring a longer recovery time, so they should be monitored very closely [104]. It has been suggested that if a patient has stable mental status and vital signs after 6 h of observation in the emergency department, he or she could be discharged unless there is some other indication for hospital admission [112].

Case reports [103] and clinical trials indicate that neostigmine or physostigmine may be helpful in the treatment of the symptoms of GHB intoxication. Physostigmine is given clinically to reverse the toxic CNS effects caused by anticholinergic agents. Three trials in humans undergoing GHB anesthesia have evaluated the use of physostigmine or neostigmine as reversal agents [118–120]. Two of these studies included the use of a neuromuscular blocker in addition to GHB, which complicate the results. In one study, effective reversal of GHB-induced sedation occurred after the administration of physostigmine alone, given intravenously as 2-mg single or repeated doses [118].

Cardiovascular Withdrawal Treatment Following GHB Overdose

Symptomatic bradycardia associated with GHB intoxication should be treated with atropine [93, 112]. However, although a single case report indicated that atropine was successful in treating a case of severe bradycardia, an approach, which has not been adequately evaluated [92].

Benzodiazepines may be given to treat the GHB withdrawal syndrome [117]. In one reported case, the withdrawal symptoms were so severe that, over a 9-day detoxification period, the patient received propranolol for cardiac stabilisation, benzodiazepines for agitation, and phenothiazines for paranoia, agitation, and delirium [108]. In another report, the patient displayed agitation, hallucinations, tachycardia, and elevated blood pressure after the cessation of GHB. Over the course of this patient's treatment, he received lorazepam 507 mg and diazepam 120 mg for agitation over a 90-h period [110]. Although benzodiazepines and other agents have been given to treat the signs and symptoms of GHB withdrawal, no standard treatment protocol exists. The treatment of GHB intoxication is mainly supportive because no

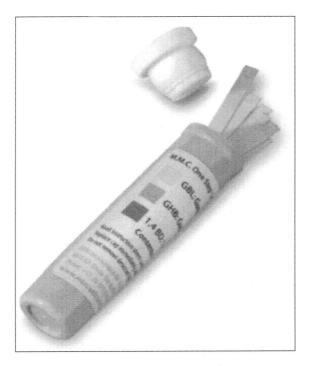

Fig. 13 The single test strip especially developed for identification of 1,4-butanediol, GBL, and GHB

specific GHB antidote has been proved effective in humans. Reduced respiratory function may require intubation or mechanical ventilation. As vomiting is a common symptom of GHB intoxication, airway protection becomes even more important to avoid the risk of aspiration. The improper manufacture of GHB can lead to a mixture of GHB and sodium hydroxide, which is very caustic and, if aspirated, is likely to cause severe damage to the lung tissue. Therefore, it is important to maintain the airway and establish intravenous access.

In order to assure the diagnosis of GBH intoxication, MMC International BC/Netherlands developed a new presumptive test for the detection of 1,4 BD (1,4-butanediol), GHB and GBL (Fig. 13). It is a strip-based colorimetric test, where the test strip is just dipped into the liquid substance while watching for a color change.

Benzylpiperazine (BZP) as a Designer Drug

The latest illegal designer drug is benzylpiperazine with trade names such as "*A2*", "*Frenzy*" and "*Nemesis*", also commonly referred to as BZP (Fig. 14). It is a recreational drug with euphoric, stimulant properties. Its dopamine and serotonin agonist mechanism of action is believed to be similar to MDMA and the effects produced by BZP are comparable to those produced by amphetamine. Adverse effects have been reported following its use including psychosis, renal toxicity, and seizures [121]. It does not appear to be very addictive and no deaths have been reported following a sole ingestion of BZP, although there have been at least two deaths from the combination of BZP and MDMA. Its sale is banned in a few countries, including the United States, Australia, New Zealand and in parts of Europe. However, its legal status is currently less restrictive in some other countries such as Ireland and Canada, although investigations and regulations are pending under European Union laws. Originally synthesized as an antihelmintic and claimed to be similar in its effect to ecstasy it has been shown to result in tachycardia, hypertonia and even epileptic seizures.

It is often claimed that BZP was originally synthesized as a potential antihelmintic (anti-parasitic) agent for use in farm animals. However, there are some references to BZP in the medical literature that predate interest in piperazines as antihelmintics. Even so, the majority of the early work with the piperazines were investigations into their potential use as antihelmintics with the earliest clinical trials in the literature relating to piperazine in the 1950s [122, 123]. It was discovered that BZP had side effects and was largely abandoned as a worm treatment. It next appears in the literature in the 1970s when it was investigated as a potential antidepressant medication, but rejected when research reported that BZP had amphetamine-like effects and was liable to abuse. The study suggested that BZP should be placed under statutory control similar to those regulating the use of amphetamine [124].

Pharmacology of Benzylpiperazine (BZP)

BZP is a piperazine derivative, which is available as either the hydrochloride salt or a free base. The hydrochloride salt is a white solid while the base form is a slightly yellowish-green liquid. BZP base is corrosive and causes burns. In countries where

Fig. 14 The molecular formula of benzylpiperazine or BZP

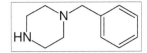

its purchase is legal, BZP products are often produced in small specialist laboratories. The raw materials can be purchased from various chemical supply agencies and formed into tablets or capsules using relatively cheap production techniques. The resulting product can be marketed at extremely high markup, so end-user prices can be as high as 300 times the bulk cost of raw ingredients. BZP is often marketed ostensibly as a *"dietary supplement"* to avoid meeting stricter laws that apply to medicines and drugs, despite the fact that BZP has no dietary value. As of late 2005 in New Zealand, the Misuse of Drugs Act ensured it can no longer be classified or marketed as a dietary supplement. Some retailers claim that BZP is a *"natural"* product, describing it as a *"pepper extract"* or *"herbal high"* when in fact the drug is entirely synthetic [125] and has not been found to occur naturally [126].

BZP has been shown to have a mixed mechanism of action, acting on the serotonergic and dopaminergic receptor systems in a similar fashion to MDMA [127]. BZP has amphetamine-like actions on the serotonin reuptake transporter, which increase serotonin concentrations in the extracellular fluids surrounding the cell and thereby increasing activation of the surrounding serotonin receptors [128, 129]. BZP has a lower potency effect on the noradrenaline reuptake transporter and the dopamine reuptake transporter [127]. BZP has a high affinity action at the alpha-2-adrenoreceptor. It is an antagonist at the receptor, like yohimbine, which inhibits negative feedback, causing an increase in released noradrenaline. BZP also acts as a non-selective serotonin receptor agonist on a wide variety of serotonin receptors [128]. Binding to $5HT_{2A}$ receptors may explain its mild hallucinogenic effects at high doses, while partial agonist or antagonist effects at the $5HT_{2B}$ receptors may explain some of BZPs peripheral side effects, as this receptor is expressed very densely in the gut, and binding to $5HT_3$ receptors may explain the common side effect of headaches, as this receptor is known to be involved in the development of migraine headaches. Hence, there is still much that is not known about the pharmacokinetics of benzylpiperazine although its metabolism is mainly through the enzymes CYP2D6 and COMT [125].

Addictive Effects of Benzylpiperazine

One in every 45 (2.2%) users of BZP in New Zealand is classed as dependent upon it, although 97.9% of users said that *"it would not be difficult to stop using legal party pills"*. About 45.2% of people who reported using both BZP and illegal drugs such as methamphetamine reported that they used BZP so that they did not have to use methamphetamine, which was perceived as more harmful [130]. Still, most of the people who use BZP, even though they say it is quite easy to stop, do not want to, and continue to use the drug, feeling that it helps them to reach higher levels of mood, sociability, and energy [130]. Studies undertaken on animals have indicated that BZP can substitute for methamphetamine in addicted rats, although it is ten times less potent and produces correspondingly weaker addictive effects [131].

The drug was classified as a Schedule I controlled substance in the United States in 2002, following a report by the DEA which incorrectly stated that BZP was 10–20 times more potent than amphetamine, when in fact BZP is ten times less potent than dexamphetamine. The DEA subsequently admitted this mistake, but nevertheless retained the Schedule 1 classification. BZP is banned in all Australian states. Victoria, the last state in which it was legal, changed its classification in September 2006. This is the date BZP and piperazine analogs became illegal in the federal schedules, which are now enacted by all Australian states and territories. BZP is also a banned substance in Japan, along with TFMPP. (3-trifluoromethylphenylpiperazine). Both Australia and Japan admit that their scheduling decisions were made primarily in response to the Schedule 1 classification given to BZP in the USA, although some instances of BZP use had been reported by law enforcement authorities in both countries. BZP is also banned in Greece, Italy, Malta, Denmark and Sweden [125].

Piperazine and salts of piperazine are classified as Prescription Only Medicines in the UK. Any products containing salts of piperazine would be licensable under the Medicines Act and consequently anyone manufacturing and supplying it legally must hold the relevant licenses to do so. BZP is not a salt of piperazine, but mislabelling of BZP products as containing *"piperazine blend"* have resulted in some prosecutions of suppliers in the UK by the Medicines and Healthcare Products Regulatory Agency, although to this date there has not been a successful prosecution in the UK for the sale of BZP. Its legal status remains uncertain. Although sale is regulated, possession of BZP is still legal. BZP and other analogs of piperazines are legal and uncontrolled in many countries such as Canada and Ireland. They are not controlled under any UN convention, so the compounds themselves are legal throughout most of the world, although in most countries their use is restricted to pharmaceutical manufacturing, and recreational use is unknown.

Benzylpiperazine is, however, subject of a European Monitoring Centre for Drugs and Drug Addiction (EMCDDA) risk assessment, the results of which will determine what, if any, control will placed on BZP throughout the European Union. The risk assessment comes about as the result of a joint Europol – EMCDDA report, which concluded that BZP needs to be looked at in more detail. The results were published in June 2007. The report concluded that the use of BZP can lead to medical problems even if the long effects are still unknown. Taking this concession as a basis, the European Commission has decided to ask the Council to place BZP under control of the UN Convention on Psychotropic Substances. On March 2008, BZP was placed under control in the EU.

Based on the recommendation of the EACD, the New Zealand government has passed legislation which placed BZP, along with the other piperazine derivatives such as TFMPP, mCPP, pFPP, MeOPP and MBZP, into Class C of the New Zealand Misuse of Drugs Act 1975. A ban was intended to come into effect in New Zealand on December 18, 2007, but the law change did not go through until the following year, and the sale of BZP and the other listed piperazines became illegal in New Zealand as of 1st of April 2008. An amnesty for possession and usage of these drugs will remain until October 2008, at which point they will become completely illegal.

Natural Products with Abuse Potential

There are several natural products such as extracts from plants, cacti, or mushrooms, which have no accepted medical use, however inherit an abuse potential

Datura a Hallucinogenic

Datura stramonium is a genus of nine species of vespertine flowering plants belonging to the family Solanaceae. Their exact natural distribution is uncertain, due to extensive cultivation and naturalization throughout the temperate and tropical regions of the globe, but is most likely restricted to the Americas, from the United States south through Mexico, where the highest species diversity occurs. Common names include Thorn Apple (from the spiny fruit), Pricklyburr (similarly), Jimson Weed, Moonflower, Hell's Bells, Devil's Weed, Devil's Cucumber, and Devil's Trumpet (from their large trumpet-shaped flowers; Fig. 15). The word datura comes from the Hindi Dhatūrā (thorn apple); record of this name dates back to 1662.

Fig. 15 Seeds and blooming of Datura stramonium as it appears in nature

Toxicity of Datura Stramonium

All Datura plants contain tropane alkaloids such as scopolamine, hyoscyamine, and atropine, primarily in their seeds and flowers. Because of the presence of these substances, Datura has been used for centuries in some cultures as a poison and hallucinogen [132, 133]. There can easily be a 5:1 variation in toxins from plant to plant, and a given plant's toxicity depends on its age, where it is growing, and local weather conditions. These wide variations make Datura exceptionally hazardous to use as a drug. In traditional cultures, users needed to have a great deal of experience and detailed plant knowledge so that no harm resulted from using it. Such knowledge is not available in modern cultures, so many incidents result from ingesting Datura. In the 1990s and 2000s, containing stories of adolescents and young adults dying or becoming seriously ill from intentionally ingesting Datura, this explains why in some parts of Europe and India, Datura has been a popular poison for suicide and murder. From 1950 to 1965, the State Chemical Laboratories in Agra investigated 2,778 deaths that were caused by ingesting Datura [132].

Due to the potent combination of anticholinergic substances it contains, Datura intoxication typically produces effects similar to that of an anticholinergic delirium: a complete inability to differentiate reality from fantasy (frank delirium, as contrasted to hallucination); hyperthermia; tachycardia; bizarre, and possibly violent behavior; and severe mydriasis with resultant painful photophobia that can last several days. Pronounced amnesia is another commonly reported effect. No other substance has received as many *"Train Wreck"* severely negative experience reports as has Datura. The overwhelming majority of those who describe the use of Datura (and to a lesser extent, Belladonna, Brugmansia and Brunfelsia) find their experiences extremely mentally and physically unpleasant and not infrequently physically dangerous.

Overdose of Datura

An overdose of Datura can occur from ingestion of as little as one half teaspoon of seeds. The overdose results in anticholinergic poisoning from tropane alkaloids, which in the worst case can lead to cardiopulmonary arrest and death. When presented with a Datura overdose, the first course of action an Emergency Room will take will be to clean out the GI tract using activated charcoal (other methods of decontamination may be contraindicated). In extreme cases (involving coma, seizures, respiratory depression, etc.) where it is known that Datura is the only substance involved, low doses of physostigmine may be utilized to reverse the life-threatening symptoms. Many hospitals will treat patients with benzodiazepines (Valium, Ativan, etc.) upon presentation of symptoms including strong hallucinations in order to calm the patient. Restraints are employed whenever a patient presents symptoms of agitation.

Patients with symptoms of anticholinergic toxicity or altered mental states are typically admitted to the ICU until such symptoms have dissipated for several hours without medications or therapies. The hallucinatory effects of Datura can last as long as 72 h, in extreme cases, equating to several days of hospitalization.

Dimethyltryptamine (DMT) a Psychedelic

DMT, also known as N,N-dimethyltryptamine, is a naturally-occurring tryptamine and potent psychedelic drug, found not only in many plants, but also in trace amounts in the human body where its natural function is undetermined. Structurally, it is analogous to the neurotransmitter serotonin and other psychedelic tryptamines such as 5-MeO-DMT and 4-HO-DMT (Fig. 16). DMT is created in small amounts by the human body during normal metabolism by the enzyme tryptamine-N-methyltransferase [134]. Many cultures, indigenous and modern, ingest DMT as a psychedelic in extracted or synthesized forms.

DMT occurs as the primary active alkaloid in several plants including such plants as Mimosa hostilis, Diplopterys cabrerana, and Psychotria viridis. DMT is found as a minor alkaloid in snuff made from Virola bark resin in which 5-MeO-DMT is the main active alkaloid [135]. DMT is also found as a minor alkaloid in the beans of Anadenanthera peregrina and Anadenanthera colubrina used to make Yopo and Vilca snuff in which bufotenin is the main active alkaloid [136].

DMT occurs naturally in many species of plants often in conjunction with its close chemical relatives 5-MeO-DMT and bufotenin (5-OH-DMT). DMT-containing plants are commonly used in several South American shamanic practices, where it is usually one of the main active constituents of the drink ayahuasca [137].

DMT is generally not active orally unless it is combined with a monoamine oxidase inhibitor (MAOI), such as harmaline. Without a MAOI, the body quickly metabolizes orally administered DMT, and it therefore has no hallucinogenic effect unless the dose exceeds monoamine oxidase's metabolic capacity. Other means of ingestion such as smoking or injecting the drug can produce powerful hallucinations and entheogenic activity for a short time (usually less than half an hour), as the DMT reaches the brain before it can be metabolised by the body's natural monoamine oxidase. Taking a MAOI prior to smoking or injecting DMT will greatly prolong and potentiate the effects of DMT. If DMT is smoked, injected, or orally ingested with a MAOI, it can produce powerful entheogenic experiences including intense visuals, euphoria, even true hallucinations (perceived extensions of reality [137]). DMT is classified in the United States as a Schedule I drug under the Controlled Substances Act of 1970.

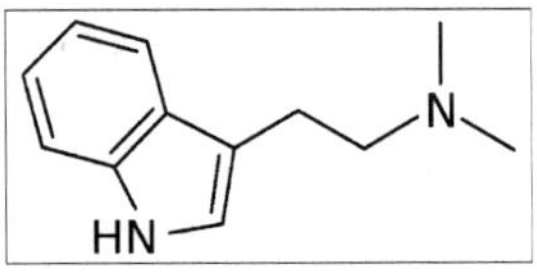

Fig. 16 Chemical structure of dimethyltryptamine (DMT; 2-(1H-indol-3-yl)-N,N-dimethyl-ethanamine)

The Mushroom Psilocybin with Psychedelic Properties

Psilocybin, the active chemical in psilocybin mushrooms can also be considered a close chemical relative of DMT, for the psilocybin molecule contains a DMT molecule at the end (4-phosphoryloxy-N,N-dimethyl-tryptamine; Fig. 17). Together with psilocin, another hallucinogenic mushroom alkaloid, it was originally isolated by Albert Hofmann and his assistant Hans Tscherter at Sandoz laboratories from magic mushrooms in 1959, guided by self-administration.

Psilocybin is a psychedelic indole of the tryptamine family, found in psilocybin mushrooms. It is present in hundreds of species of fungi, including those of the genus Psilocybe, such as Psilocybe cubensis and Psilocybe semilanceata, but also reportedly isolated from a dozen or so other genera. Psilocybin mushrooms are commonly called *"magic mushrooms"* or more simply *"shrooms"* (Fig. 18).

Toxicity of Psilocybinly

The toxicity of psilocybin is relatively low; in rats, the oral LD_{50} is 280 mg/kg, approximately one and a half times that of caffeine. When administered intravenously in rabbits, psilocybin's LD_{50} is approximately 12.5 mg/kg [138]. However, rabbits are extremely intolerant to the effects of most psychoactive drugs. The lethal dose from psilocybin intake alone is unknown at recreational or medicinal levels, and has never been documented; psilocybin makes up roughly 1% of the weight of Psilocybe cubensis mushrooms, and so nearly 1.7 kg of dried mushrooms, or 17 kg of fresh mushrooms, would be required for a 60 kg person to reach the 280 mg/kg LD_{50} rate of rats.

Pharmacological Effects of Psilocybinly

The effects of psilocybin are highly variable, and dependent on the current mood and overall sense of well-being by the individual. Initially the subject may begin to feel somewhat disorientated, lethargic, and euphoric or sometimes depressed. At

Fig. 17 Psilocybin, or O-phosphoryl-4-hydroxy-N,N-dimethyltryptamine, the active chemical compound in hallucinogenic (magic) mushrooms

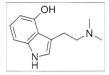

Fig. 18 Appearance of psilocybin mushrooms inducing psychedelic properties

low doses, hallucinatory effects may occur, including enhancement of colors and the animation of geometric shapes. Closed-eye hallucination may occur, where the affected individual may see multi-coloured geometric shapes and vivid imaginative sequences. At higher doses, hallucinatory effects increase and experiences tend to be less social and more introspectic or entheogenic [138]. Open-eye visuals are more common, and may be very detailed although rarely confused with reality.

Distortions in the experience of time in psilocybin-induced states have been subjectively reported, and objectively measured [139, 140]. In these studies, psilocybin significantly decreased subjects' reproduction of time intervals longer than 2.5 s, impaired their ability to synchronize to inter-beat intervals longer than 2 s, and reduced their preferred tapping rate. Recent studies into the effects of psilocybin on time interval reproduction may shed light on qualitative alterations of time experience in experimentally-induced altered states of consciousness, mystical states, or in psychopathology [141].

Users having a pleasant experience can feel ecstatic, a sense of connection to others, nature, the universe, and other feelings/emotions are often intensified. Difficult experiences or bad trips occur due to a variety of reasons. Tripping during an emotional/physical low, or in a non-supportive/inadequate/etc. environment (see: set and setting) could possibly cause anxiety or some sort of freak-out. Latent psychological issues may be triggered by the strong emotional components of the experience [142], Some of these individuals report that they have experienced a *"spiritual episode"*. For example, in the Marsh Chapel Experiment, which was run by a graduate student at Harvard Divinity School under the supervision of Timothy Leary, almost all of the graduate degree divinity student volunteers who received psilocybin reported profound religious experiences.

Psilocybin and its congener psilocin are listed as Schedule I drugs under the United Nations 1971 Convention on Psychotropic Substances. Possession, and in some cases usage, of psilocybin or psilocin has been outlawed in most countries across the globe.

Ibogaine, Psychedelic Molecule with Anti-addictive Properties

Ibogaine is a naturally occurring psychoactive compound found in a number of plants, principally in a member of the dogbane family known as iboga (Tabernanthe iboga; Fig. 19). Bark Ibogaine-containing preparations are used in medicinal and ritual purposes by African spiritual traditions of the Bwiti, who claim to have learned it from the Pygmy. In recent times, it has been identified as having anti-addictive properties. Ibogaine is an indole alkaloid that is obtained either by extraction from the iboga plant or by semi-synthesis from the precursor compounds voacangine, another plant alkaloid. Although a full organic synthesis of ibogaine has been achieved but is too expensive and challenging to produce any commercially significant yield.

In the early 1960s, ibogaine was accidentally discovered to cause sudden and complete interruption of heroin addiction without withdrawal in a matter of hours. Since that time, it has been the subject of scientific investigation into its abilities to interrupt addictions to heroin, alcohol, and cocaine. Anecdotal reports also suggest that ibogaine may have potential to drive introspection that helps elucidate the psychological issues and behavior patterns that drive addictions or other problems. However, ibogaine therapy for drug addiction is the subject of some controversy. Due to its hallucinogenic properties, it has been placed in the strictest drug prohibition schedules in the United States and a handful of other countries. Canada and Mexico both allow ibogaine treatment clinics to operate and openly contribute to further understanding of the addictive process.

While ibogaine's prohibition has slowed scientific research into its anti-addictive properties, the use of ibogaine for drug treatment has grown in the form of a large worldwide medical subculture [143]. Ibogaine is now used by treatment clinics in 12 countries on six continents to treat addictions to heroin, alcohol, powder cocaine, crack cocaine, and methamphetamine, as well as to facilitate psychological introspection and spiritual exploration

Fig. 19 Chemical structure of ibogaine (12-methoxyibogamine) with preported anti-drug addictive properties

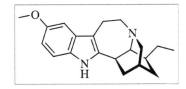

Pharmacology of Ibogaine

At doses of around 3–5 mg/kg of body weight, ibogaine has a mild stimulant effect. The high-dose ibogaine experience of 10 mg/kg or greater most commonly occurs as two distinct phases: the visual phase and the introspective phase. The visual phase is characterized by open-eye visuals, closed-eye visuals, and dreamlike sequences. Objects may be seen as distorted, projecting tracers, or having moving colors or textures. With the eyes closed, extremely detailed and vivid geometric and fractal visions may be seen. Subjective reports often include a movie-like recollection of earlier life experiences as well as dreamlike sequences with symbolism of one's present or anticipated future. Other effects in the visionary phase may include laughing, sensations of euphoria or fear, and temporary short-term memory impairment. The visionary phase usually ends after 1–4 h, after which the introspective phase begins [144, 145].

The introspective phase is typically reported to bring elevated mood, a sense of calm and euphoria, and a distinct intellectual and emotional clarity. Subjects often report being able to accomplish deep emotional and intellectual introspection into psychological and emotional concerns. It is also during this period that opioid addicts first notice the absence of withdrawal cravings. The duration of the introspective phase is highly variable, usually lasting hours but sometimes lasting days. Alper [146] Mash and coworkers [147] published data demonstrating ibogaine's efficacy in attenuating opioid withdrawal in drug-dependent human subjects. As to its mode of action ibogaine is a weak $5HT_{2A}$ receptor agonist [148], and although it is unclear how significant this action is for the anti-addictive effects of ibogaine, it is likely to be important for the hallucinogenic effects. [149]. In this regard it has been shown to be a sigma2 receptor agonist [150].

Ibogaine and its salts were regulated by the US Food and Drug Administration in 1967 pursuant to its enhanced authority to regulate stimulants, depressants, and hallucinogens granted by the 1965 Drug Abuse Control Amendments (DACA) to the Federal Food, Drug, and Cosmetic Act. In 1970, with the passage of the Controlled Substances Act, it was classified as a Schedule I-controlled substance in the United States, along with other psychedelics such as LSD and mescaline. Since that time, several other countries, including Sweden, Denmark, Belgium, and Switzerland, have also banned the sale and possession of ibogaine. Although illegal, ibogaine has been used by hundreds of drug addicts in the United States and abroad [151].

Peyote, a Mescaline-Containing Cactus

The naturally occurring psychedelic alkaloid of the phenethylamine class is mainly used as a recreational drug, an entheogen, and a tool to supplement various practices for transcendence, including in meditation, psychonautics, art projects, and psychedelic psychotherapy. It occurs naturally in the peyote cactus (Lophophora williamsii), the San Pedro cactus (Echinopsis pachanoi) and the Peruvian Torch cactus (Echinopsis peruviana; Fig. 20), and in a number of other members of the Cactaceae.

Peyote or Lophophora williamsii is a small, spineless cactus. It is native to southwestern Texas, through central Mexico. It is found primarily in the Chihuahuan desert and in the states of Tamaulipas and San Luis Potosi among scrub, especially where there is limestone. It is well known for its psychoactive alkaloids particularly mescaline. It is currently used world wide as a recreational drug, an entheogen, and supplement to various transcendence practices including meditation, psychonautics, and psychedelic psychotherapy. Peyote has a long history of ritual religious and medicinal use by indigenous Americans.

Behavioral Effects Produced by Mescaline

Mescaline, being the active ingredient of Peyote, was first isolated and identified in 1897 by the German Arthur Heffter and first synthesized in 1919 by Ernst Späth (Fig. 21).

The hallucinations produced by mescaline are somewhat different from those of LSD. Hallucinations are consistent but are typically intensifications of the stimulus properties of objects and sounds. Prominence of color is distinctive, appearing brilliant and intense. Placing a strobing light in front of closed eyelids can produce brilliant visual effects at the peak of the experience. Recurring visual patterns observed during the mescaline experience include stripes, checkerboards, angular spikes, multicolored dots, and very simple fractals, which turn very complex. Like LSD, mescaline induces distortions of form and kaleidoscopic experiences but which manifest more clearly with eyes closed and under low lighting conditions; however, all of these visual descriptions are purely subjective. And like with LSD, synesthesia can occur especially with the help of music [152]. An unusual but

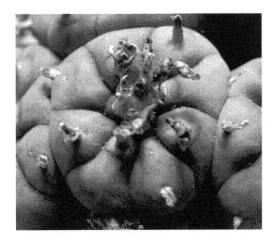

Fig. 20 Appearance of the naturally peyote cactus (Lophophora williamsii), the Peruvian Torch cactus (Echinopsis peruviana)

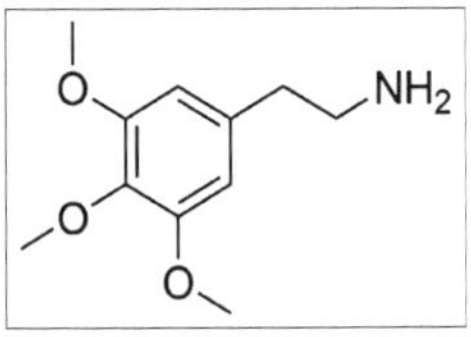

Fig. 21 Chemical structure of mescaline or 3,4,5-trimethoxy-phenethylamine, the primary psychoactive compound in peyote

unique characteristic of mescaline use is the "*geometricization*" of three-dimensional objects. The object can appear flattened and distorted, similar to the presentation of a Cubist painting. Similarly to other psychedelic agents, mescaline binds to, and activates the serotonin $5HT_{2A}$ receptor with a high nanomolar affinity [153]. Mescaline elicits a pattern of sympathetic arousal, with the peripheral nervous system being a major target for this drug, where the effects last for up to 12 h [154].

Peyote is listed by the United States DEA as a Schedule I controlled substance. Although many American jurisdictions specifically allow religious use of peyote, religious or therapeutic use not under the aegis of the Native American Church has often been targeted by local law enforcement agencies, and non-Natives attempting to establish spiritual centers based on the consumption of peyote as a sacrament or as medicine, such as the Peyote Foundation in Arizona, have been prosecuted. While in Canada the active ingredient of Peyote Mescaline is listed as a Schedule III controlled substance under the Canadian Controlled Drugs and Substances Act, peyote is specifically exempt. Internationally, Article 32 of the Convention on Psychotropic Substances allows nations to exempt certain traditional uses of peyote from prohibition.

LSD, a Semisynthetic Psychedelic Drug

Lysergic acid diethylamide, LSD, LSD-25, or *"acid"*, is a semisynthetic psychedelic drug of the ergoline family. Its unusual psychological effects, which include visuals of colored patterns behind the eyes, a sense of time distorting, and crawling geometric patterns, has made it one of the most widely known psychedelic drugs. It has been used mainly as a recreational drug, an entheogen, and as a tool to supplement various practices for transcendence, including in meditation, psychonautics, art projects, and illicit (formerly legal) psychedelic therapy. Formally, LSD is classified as a hallucinogen of the psychedelic type.

It was synthesized from lysergic acid derived from ergot, a grain fungus that typically grows on rye, and was first synthesized by the Swiss chemist Albert Hofmann (Fig. 22). The short form LSD comes from its early code name LSD-25, which is an abbreviation for the German *"Lysergsäure-diethylamid"* followed by a sequential number of synthetic agents.

Pharmacology of LSD

The inventor who had tried it on himself gives the best description of the effects of LSD at a dose of 250 µg. While riding a bike he had the sensation of being stationary, unable to move from where he was, despite the fact that he was moving very rapidly. Having no abnormal physical symptoms other than extremely dilated pupils. He was terrified that his body had been possessed by a demon, that his next door neighbor was a witch, and that his furniture was threatening him, feared he had become completely insane. At this time Hofmann said that the feelings of fear had started to give way to feelings of good fortune and gratitude, and that he was now enjoying the colors and plays of shapes that persisted behind his closed eyes. Hofmann mentions seeing *"fantastic images"* surging past him, alternating and opening and closing themselves into circles and spirals and finally exploding into colored fountains and then rearranging themself in a constant flux. Hofmann mentions that during the condition every acoustic perception, such as the sound of a passing automobile, was transformed into optical perceptions. Eventually after a refreshing sleep he felt

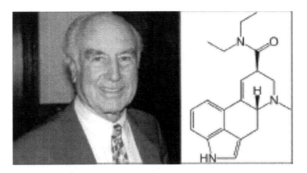

Fig. 22 The inventor of lysergic acid diethylamide (*right*), the Swiss chemist Albert Hoffmann (*left*), who searching for medically useful ergot alkaloid derivatives, synthesized LSD-25 on November 16, 1938 at the Sandoz Laboratories in Basel/Switzerland

refreshed and clearheaded, though somewhat physically tired. He also stated that he had a sensation of well being and renewed life and that his breakfast tasted unusually delicious. Upon walking in his garden he remarked that all of his senses were "*vibrating in a condition of highest sensitivity, which then persisted for the entire day*".

Research with LSD

Since LSD could produce changes in perceptions and emotions, early researchers hypothesized that the cause of some mental illnesses, particularly schizophrenia, were caused by endogenous compounds with a similar activity to LSD [12]. Much of the research during the late 1940s dealt with this hypothesis and many LSD sessions conducted for scientific study were often termed "experimental psychoses", and this is where the terms "*psychoactive*", "*psychotomimetic*" and "*hallucinogenic*" were coined to refer to such drugs.

In 1953 and 1954, scientists working for MI6 (British secret service) dosed servicemen in an effort to find a "truth drug" that could be used in interrogations. The test subjects were not informed that they were being given LSD, and had in fact been told that they were participating in a medical project to find a cure for the common cold. Also, during the Cold War, intelligence agencies (CIA) were keenly interested in the possibilities of using LSD for interrogation and mind control, as well as for large-scale social engineering. The CIA research on LSD, most of which was done under Project MKULTRA, the code name for a CIA mind-control research program, began in the 1950s and continued until the late 1960s.

LSD is being consumed in drinks or is soaked up on blotting paper (*paper-trip*), which is placed on the tongue (Fig. 23) from where the active compound is being absorbed into the blood stream (50–200 µg) resulting in psychedelic visions. It is now hypothesised that LSD could be used in the treatment of obsessive-compulsive

Fig. 23 Colorful printed blotting paper with soaked LSD as it is being sold illegally

disorder and severe depression [155] due to the substance's stimulation of the $5HT_{2A}$ receptors [156]. There has been additional interest in studying the effects of LSD on cluster headaches [157], although the current status of this research is uncertain.

The Legal Status of LSD

LSD is a Schedule I in the United States. This means it is illegal to manufacture, buy, possess, process or distribute LSD without a DEA license. The United Nations Convention on Psychotropic Substances (adopted in 1971) requires its parties to prohibit LSD. Hence, it is illegal in all parties to the convention, which includes the United States, Australia, and most of Europe. However, enforcement of extant laws varies from country to country.

5-MeO-DIPT, a Psychedelic Structurally Related to Psiloc(yb)in

5-MeO-DIPT, also known as 5-methoxy-N,N-diisopropyltryptamine (Fig. 24), *foxy methoxy*, now often known by its recently coined name *"foxy"*, is a stimulating psychedelic tryptamine publicized as an erotic enhancer. It experienced a surge in popularity due to Internet sales from 1999 to 2001, before being made illegal in the US in 2003. It is orally active in dosages ranging between 6 and 20 mg. Being a pschoactive tryptamine analog, it was originally developed by Alexander Shulgin around 1980. Reactions to 5-MeO-DIPT vary dramatically from those who find it compelling, sexy, exhilerating, interesting, or joyful to those who find it nauseating, irritating, diarrhea-inducing, and generally unpleasant. It is available primarily in powder form, though it is also found in liquid and pressed tablets, and is almost always taken orally, though sometimes snorted. Many users note an unpleasant body load accompanies higher dosages. 5-MeO-DIPT is also taken by insufflation, or sometimes it is smoked or injected. Some users also report sound distortion, also noted with the related drug, DIPT.

Pharmacology of 5-MeO-DIPT

The effects of 5-MeO-DIPT can be quite stimulating and is known for causing mood lift, euphoria, intensification of tactile sensations, smell, together with emotionally opening. It also gained a reputation for being a sex-drug after an early report described a sensual experience and a lurid article described it in PlayBoy. It is also well known for causing diarrhea and gas, although not in all users.

Though the peak effects are not exceedingly long, the lingering stimulation that can last for up to 12 h (or more) after the peak has given it a reputation for being quite long. The after effects often include tense muscles and can sometimes include feelings of anxiety or inability to relax.

Fig. 24 Chemical structure of *Foxy*, a N,N-diisopropyl-5-methoxy-tryptamine

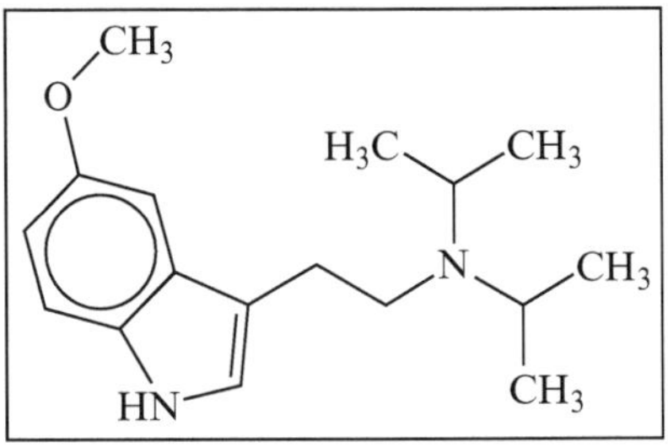

Legal Status of 5-MeO-DIPT

5-MeO-DIPT has been illegal in Germany since September 1999, Greece since February 2003, Denmark since February 2004, Sweden since October 2004, Japan since April 2005, and Singapore since early 2006. On April 4, 2003, the United States DEA added 5-MeO-DIPT and its analogs to Schedule I of the Controlled Substances Act under "emergency scheduling" procedures. The drugs were officially placed into Schedule I on September 29, 2004.

References

1. Langston JW, Ballard P, Tetrud JW, Irwin I. Chronic parkinsonism in humans due to a product of meperidine-analog synthesis. Science. 1983;219:979–80.
2. Langston JW, Irwin I, Langston EB, Forno LS. Pargyline prevents MPTP-induced parkinsonism in primates. Science. 1984;225:1480–2.
3. Shulgin A, Shulgin A. Pihkal: a chemical love story. Berkeley, CA: Transform Press; 1995.
4. Shulgin A, Shulgin A. Thikal: the continuation. Berkeley, CA: Transform Press; 1997.
5. Tunnicliff G. Significance of γ-hydroxybutyric acid in the brain. Gen Pharmacol. 1992;23:1027–34.
6. Castelli MP, Ferraro L, Mocci I, Carta F, Carai MA, Antonelli T, et al. Selective gamma-hydroxybutyric acid receptor ligands increase extracellular glutamate in the hippocampus, but fail to activate G protein and to produce the sedative/hypnotic effect of gamma-hydroxybutyric acid. J Neurochem. 2000;87:722–32.
7. Spanos PF, Tagliamonte A, Tagliamonte P, Gessa GL. Stimulation of brain dopamine synthesis by γ-hydroxybutyrate. J Neurochem. 1971;18:1831–6.
8. Feigenbaum JJ, Howard SG. Does gamma-hydroxybutyrate inhibit or stimulate central DA release? Int J Neurosci. 1996;88:53–69.
9. Tunnicliff G. Sites of action of gamma-hydroxybutyrate (GHB) – a neuroactive drug with abuse potential. J Toxicol Clin Toxicol. 1997;35:581–90.
10. Andriamampandry C, Taleb O, Viry S, Muller C, Humbert J, Gobaille S, et al. Cloning and characterization of a rat brain receptor that binds the endogenous neuromodulator gamma hydroxybutyrate (GHB). FASEB. 2003;17:1691–3.
11. Hechler V, Gobaille S, Maitre M. Selective distribution pattern of gamma-hydroxybutyrate receptors in the rat forebrain and midbrain as revealed by quantitative autoradiography. Brain Res. 1992;572:345–8.
11a. Colombo G, Agabio R, Bourguignon J, Faddo F, Cobina C, Maitre M, et al. Blockade of the discriminative stimulus effects of gamma-hydroxybutyic acid (GHB) by the GHB receptor autagonist NCS-382. Physiol Behav. 1995;58:587–590.
12. Laborit H. Sodium 4-hydroxybutyrate. Int J Neuropharmacol. 1964;3:433–52.
13. Bessman SP, Fishbein WN. Gamma-hydroxybutyrate, a normal brain metabolite. Nature. 1963;200:1207–8.
14. Root B. Oral premedication of children with 4-hydroxybutyrate. Anesthesiology. 1965;26:259–60.
15. Winters WD, Spooner CE. Various seizure activities following gamma-hydroxybutyrate. Int J Neuropharmacol. 1965;4:197–200.
16. Godschalk M, Dzoljic MR, Bonta IL. Slow wave sleep and a state resembling absence epilepsy induced in the rat by γ-hydroxybutyrate. Eur J Pharmacol. 1977;44:105–11.
17. Solway J, Sadove MS. 4-Hydroxybutyrate: a clinical study. Anesth Analg. 1965;44:532–9.
18. Mamelak M, Scharf MB, Woods M. Treatment of narcolepsy with gamma-hydroxybutyrate: a review of clinical and sleep laboratory findings. Sleep. 1986;9:285–9.

19. Takahara J, Yunoki S, Yakushiji W, Yamauchi J, Yamane Y, Ofuji T. Stimulatory effects of gamma-hydroxybutyric acid on growth hormone and prolactin release in humans. J Clin Endocrinol Metab. 1977;44:1014–7.
20. Gallimberti L, Canton G, Gentile N, et al. Gamma-hydroxybutyric acid for treatment of alcohol withdrawal syndrome. Lancet. 1989;2:787–9.
21. Gallimberti L, Cibin M, Pagnin P, et al. Gamma-hydroxybutyric acid for treatment of opiate withdrawal syndrome. Neuropsychopharmacology. 1993;9:77–81.
22. WHO questionnaire for review of dependence-producing psychoactive substances, 2000.
23. Sanguineti VR, Angelo A, Frank MR. GHB: a home brew. Am J Drug Alcohol Abuse. 1997;23: 634–42.
24. Wiley J, Dick R, Arnold T. Hematuria from home-manufactured GHB. J Toxicol Clin Toxicol. 1998;46:502–3.
25. Henretig F, Vassalluzo C, Osterhoudt K, et al. "Rave by net": gamma-hydroxybutyrate (GHB) toxicity from kits sold to minors via the internet [abstr]. J Toxicol Clin Toxicol. 1998;36:503.
26. Palatini P, Tedeschi L, Frison G, et al. Dose-dependent absorption and elimination of gamma-hydroxybutyric acid in healthy volunteers. Eur J Clin Pharmacol. 1993;45:353–6.
27. Ferrara SD, Zotti S, Tedeschi L, et al. Pharmacokinetics of gamma-hydroxybutyric acid in alcohol dependent patients after single and repeated oral doses. Br J Clin Pharmacol. 1992;34:231–5.
28. Scharf MB, Lai AA, Branigan B, Stover R, Berkowitz DB. Pharmacokinetics of gamma-hydroxybutyrate (GHB) in narcoleptic patients. Sleep. 1998;21:507–14.
29. Helrich M, McAslan TC, Skolnik S, Bessman SP. Correlation of blood levels of 4-hydroxybutyrate with state of consciousness. Anesthesiology. 1964;25:771–5.
30. Hoes MJ, Vree TB, Guelen PJ. Gamma-hydroxybutyric acid as hypnotic. Encephale. 1980;6:93–9.
31. Vickers MD. Gamma-hydroxybutyric acid. Int Anesthesiol Clin. 1969;7:75–89.
32. Lettieri J, Fung H. Absorption and first-pass metabolism of (14)C-gamma-hydroxybutyric acid. Res Commun Chem Pathol Pharmacol. 1976;13:435–7.
33. Roth RH, Giarman NJ. Conversion in vivo of gamma-aminobutyric to γ-hydroxybutyric acid in the rat. Biochem Pharmacol. 1969;18:247–50.
34. Cash CD. Gamma-hydroxybutyrate: an overview of the pros and cons for it being a neurotransmitter and/or a useful therapeutic agent. Neurosci Biobehav Rev. 1994;18:291–304.
35. Mandel P, Maitre M, Vayer P, Hechler V. Function of -hydroxybutyrate: a putative neurotransmitter. Biochem Soc Trans. 1987;15:215–7.
36. Vayer P, Mandel P, Maitre M. Gamma-hydroxybutyrate, a possible neurotransmitter. Life Sci. 1987;41:1547–58.
37. Tunnicliff G. Significance of gamma-hydroxybutyric acid in the brain. Gen Pharmacol. 1992;23:1027–34.
38. Tunnicliff G. Sites of action of gamma-hydroxybutyrate (GHB): a neuroactive drug with abuse potential. J Toxicol Clin Toxicol. 1997;35:581–90.
39. Anderson RA, Ritzmann RF, Tabakoff B. Formation of gamma-hydroxybutyrate in brain. J Neurochem. 1977;28:633–9.
40. Doherty JD, Roth RH. Metabolism of gamma-hydroxy-[1–14C] butyrate by rat brain: relationship to the Krebs cycle and metabolic compartmentation of amino acids. J Neurochem. 1978;30:1305–9.
41. Roth RH, Doherty JD, Walters JR. Gamma-hydroxybutyrate: a role in the regulation of central dopaminergic neurons ? Brain Res. 1980;189:556–60.
42. Vayer P, Ehrhardt J, Gobaille S, Mandel P, Maitre M. Gamma-hydroxybutyrate distribution and turnover rates in discrete brain regions of the rat. Neurochem Int. 1988;12:53–9.
43. Benavides J, Rumigny J, Bourguignon JJ, et al. High-affinity binding site for gamma-hydroxybutyric acid in rat brain. Life Sci. 1982;30:953–61.
44. Hechler V, Bourguignon JJ, Wermuth CG, Mandel P, Maitre M. Gamma-hydroxybutyrate uptake by rat brain striatal slices. Neurochem Res. 1985;10:387–96.
45. Nelson T, Kaufman E, Kline J, Sokoloff L. The extraneural distribution of γ-hydroxybutyrate. J Neurochem. 1981;37:1345–8.

46. Vayer P, Maitre M. Gamma-hydroxybutyrate stimulation of the formation of cyclic GMP and inositol phosphates in rat hippocampal slices. J Neurochem. 1989;52:1382–7.
47. Maitre M, Hechler V, Vayer P, et al. A specific gamma-hydroxybutyrate receptor ligand possesses both antagonistic and anticonvulsant properties. J Pharmacol Expt Ther. 1990;255:657–63.
48. Bernasconi R, Lauber J, Marescaux C, et al. Experimental absence seizures: potential role of γ-hydroxybutyric acid and GABAB receptors. J Neural Transm Suppl. 1992;35:155–77.
49. Xie X, Smart TG. Gamma-hydroxybutyrate hyperpolarizes hippocampal neurones by activating GABAB receptors. Eur J Pharmacol. 1992;219:292–4.
50. Serra M, Sanna E, Foddi C, Concas A, Biggio G. Failure of gamma-hydroxybutyrate to alter the function of the GABAA receptor complex in the rat cerebral cortex. Psychopharmacology. 1991;104:351–5.
51. Gessa GL, Vargiu L, Crabai F, Boero GC, Caboni F, Camba R. Selective increase of brain dopamine induced by gamma-hydroxybutyrate. Life Sci. 1966;5:1921–30.
52. Bustos G, Roth RH. Effect of gamma-hydroxybutyrate on the release of monoamines from the rat striatum. Br J Pharmacol. 1972;44:817–20.
53. Hechler V, Gobaille S, Bourguignon JJ, Maitre M. Extracellular events induced by gamma-hydroxybutyrate in striatum: a microdialysis study. J Neurochem. 1991;56:938–44.
54. Snead OCI, Bearden LJ. Naloxone overcomes the dopaminergic, EEG, and behavioral effects of gamma-hydroxybutyrate. Neurology. 1980;30:832–8.
55. Grove-White IG, Kelman GR. Effect of methohexitone, diazepam and sodium 4-hydroxybutyrate on short-term memory. Br J Anaesth. 1971;43:113–6.
56. Mamelak M, Sowden K. The effect of gamma-hydroxybutyrate on the H-reflex: pilot study. Neurology. 1983;33:1497–14500.
57. Lapierre O, Lamarre M, Montplaisir J, Lapierre G. The effect of γ-hydroxybutyrate: a double-blind study of normal subjects [abstr]. Sleep Res. 1988;17:99.
58. Metcalf DR, Emde RN, Stripe JT. An EEG-behavioral study of sodium hydroxybutyrate in humans. Electroenceph Clin Neurophysiol. 1966;20:506–12.
59. Prevention CfDCa. Multistate outbreak of poisonings associated with illicit use of γ-hydroxybutyrate. MMWR. 1990;39:861–3.
60. Virtue RW, Lund LO, Beckwitt HJ, Vogel JH. Cardiovascular reactions to gamma-hydroxybutyrate in man. Can Anaesth Soc J. 1966;13:119–23.
61. Appleton PJ, Burn JM. A neuroinhibitory substance: gamma-hydroxybutyric acid. Anseth Analg. 1968;47:164–70.
62. Van Cauter E, Plat L, Scharf MB, Leproult R, Cespedes S, L'Hermite-Baleriaux M. Simultaneous stimulation of slow-wave sleep and growth hormone secretion by gamma-hydroxy-butyrate in normal young men. J Clin Invest. 1997;100:745–53.
63. Broughton R, Mamelak M. Effects of nocturnal gamma-hydroxybutyrate on sleep/waking patterns in narcolepsy-cataplexy. Can J Neurolo Sci. 1980;7:23–31.
64. Scrima L, Hartman PG, Johnson FHJ, Thomas EE, Hiller FC. The effects of gamma-hydroxybutyrate on the sleep of narcolepsy patients: a double-blind study. Sleep. 1990;13:479–90.
65. Scharf MB, Brown D, Woods M, Brown L, Hirschowitz J. The effects and effectiveness of gamma-hydroxybutyrate in patients with narcolepsy. J Clin Psychiatry. 1985;46:222–5.
66. Broughton R, Mamelak M. Effects of nocturnal gamma-hydroxybutyrate on sleep/waking patterns in narcolepsy-cataplexy. Can J Neurolo Sci. 1980;7:23–31.
67. Lapierre O, Montplaisir J, Lamarre M, Bedard MA. The effect of gamma-hydroxybutyrate on nocturnal and diurnal sleep of normal subjects: further considerations on REM sleep-triggering mechanisms. Sleep. 1990;13:24–39.
68. Entholzner E, Mielke L, Pichlmeier R, Weber F, Schneck H. EEG changes during sedation with γ-hydroxybutyric acid. Anaesthesist. 1995;44:345–50.
69. Kleinschmidt S, Schellhase C, Mertzlufft F. Continuous sedation during spinal anaesthesia: gamma-hydroxybutyrate vs propofol. Eur J Anaesthesiol. 1999;16:23–30.
70. Mamelak M. Gamma-hydroxybutyrate: an endogenous regulator of energy metabolism. Neurosci Biobehav Rev. 1989;13:187–98.

71. Li J, Stokes SA, Woeckener A. A tale of novel intoxication: a review of the effects of γ-hydroxybutyric acid with recommendations for management. Ann Emerg Med. 1998;31:729–36.
72. Broughton R, Mamelak M. The treatment of narcolepsy-cataplexy with nocturnal gamma-hydroxybutyrate. Can J Neurol Sci. 1979;6:1–6.
73. Gessa GL, Addolorato G, Caputo F, et al. Symposium on gamma-hydroxybutyric acid (GHB): a neurotransmitter, a medicine, a drug of abuse. Alcohol. 2000;20:213–304.
74. Addolorato G, Caputo F, Stefanini GF, Gasbarrini G. Gamma-hydroxybutyric acid in the treatment of alcohol dependence: possible craving development for the drug. Addiction. 1997;92:1035–6.
75. Bessman SP, Rossen J, Layne EC. Gamma-aminobutyric acid-glutamic acid transamination in brain. J Biol Chem. 1953;201:385–91.
76. Fishbein WN, Bessman SP. Gamma-hydroxybutyrate in mammalian brain. Reversible oxidation by lactic dehydrogenase. J Biol Chem. 1964;239:357–61.
77. Nelson T, Kaufman EE. Developmental time courses in the brain and kidney of two enzymes that oxidize -hydroxybutyrate. Dev Neurosci. 1994;16:352–8.
78. Dyer JE. Gamma-hydroxybutyrate: a health-food product producing coma and seizure-like activity. Am J Emerg Med. 1991;9:321–4.
79. Galloway GP, Frederick SL, Staggers FEJ, Gonzales M, Stalcup SA, Smith DE. Gamma-hydroxybutyrate: an emerging drug of abuse that causes physical dependence. Addiction. 1997;92:89–96.
80. Chin MY, Kreutzer RA, Dyer JE. Acute poisoning from gamma-hydroxybutyrate in California. West J Med. 1992;156:380–4.
81. McDowell DM. MDMA, ketamine, GHB, and the "club drug" scene. In: Galanter M, Kleber HD, editors. Textbook of substance abuse treatment. Washington, DC: American Psychiatric Press; 1999. pp. 295–305.
82. McCabe ER, Layne EC, Sayler DF, Slusher N, Bessman SP. Synergy of ethanol and a natural soporific: gamma-hydroxybutyrate. Science. 1971;171:404–6.
83. Gerra G, Caccavari R, Fontanesi B, et al. Flumazenil effects on growth hormone response to γ-hydroxybutyric acid. Int Clin Psychopharmacol. 1994;9:211–5.
84. Addolorato G, Capristo E, Gessa GL, Caputo F, Stefanini GF, Gasbarrini G. Long-term administration of GHB does not affect muscular mass in alcoholics. Life Sci. 1999;65:191–5.
85. Centers for Disease Control and Prevention. Multistate outbreak of poisonings associated with illicit use of gamma-hydroxybutyrate. MMWR. 1990;39:861–3.
86. Centers for Disease Control and Prevention. Gamma-hydroxybutyrate use: New York and Texas, 1995–1996. MMWR. 1997;46:281–3.
87. Ross TM. Gamma-hydroxybutyrate overdose: two cases illustrate the unique aspects of this dangerous drug. J Emerg Nurs. 1995;21:374–6.
88. Thomas G, Bonner S, Gascoigne A. Coma induced by abuse of gamma-hydroxybutyrate (GHB or "Liquid Ecstasy"): a case report. BMJ. 1997;314:35–6.
89. Steele MT, Watson WA. Acute poisoning from gamma-hydroxybutyrate (GHB). Mol Med. 1995;92:354–47.
90. Libetta C. Gamma-hydroxybutyrate poisoning. J Accid Emerg Med. 1997;14:411–2.
91. Ryan JM, Stell I. Gamma-hydroxybutyrate: a coma- inducing recreational drug. J Accid Emerg Med. 1997;14:259–61.
92. Chin RL, Sporer KA, Cullison B, Dyer JE, Wu TD. Clinical course of gamma-hydroxybutyrate overdose. Ann Emerg Med. 1998;31:716–22.
93. Viera AJ, Yates SW. Toxic ingestion of gamma-hydroxybutyric acid. South Med J. 1999;92:404–5.
94. Louagie HK, Verstraete AG, De Soete CJ, Baetens DG, Calle PA. A sudden awakening from a near coma after combined in-take of γ-hydroxybutyric acid (GHB) and ethanol. J Toxicol Clin Toxicol. 1997;35:591–4.
95. Li J, Stokes SA, Woeckener A. A tale of novel intoxication: seven cases of gamma-hydroxybutyric acid overdose. Ann Emerg Med. 1998;31:723–8.
96. Ingels M, Rangan C, Bellezzo J, Clark RF. Coma and respiratory depression following the ingestion of GHB and its precursors: three cases. J Emerg Med. 2000;19:47–50.

97. Food and Drug Administration. FDA warns about products containing gamma-butyrolactone or GBL and asks companies to issue a recall. Rockville, MD: National Press Office; 1999.
98. Garrison G, Muller P. Clinical features and outcomes after unintentional gamma-hydroxybutyrate (GHB) overdose [abstr]. J Toxicol Clin Toxicol. 1998;36:503–4.
99. Louagie HK, Verstraete AG, De Soete CJ, Baetens DG, Calle PA. A sudden awakening from a near coma after combined in-take of gamma-hydroxybutyric acid (GHB) and ethanol. J Toxicol Clin Toxicol. 1997;35:591–4.
100. Ferrara SD, Tedeschi L, Frison G, Rossi A. Fatality due to gamma-hydroxybutyric acid (GHB) and heroin intoxication. J Forensic Sci. 1995;40:501–4.
101. Fieler EL, Coleman DE, Baselt RC. Gamma-hydroxybutyrate concentrations in pre- and postmortem blood and urine. Clin Chem. 1998;44:692–3.
102. Timby N, Eriksson A, Bostrom K. Gamma-hydroxybutyrate-associated deaths. Am J Med. 2000;108:418–519.
103. Yates SW, Viera AJ. Physostigmine in the treatment of gamma-hydroxybutyric acid overdose. Mayo Clin Proc. 2000;75:401–2.
104. Hodges B, Everett J. Acute toxicity from home-brewed gamma-hydroxybutyrate. J Am Board Fam Pract. 1998;11:154–7.
105. Brown TC. Gamma-hydroxybutyrate in pediatric anaesthesia. Aust N Z J Surg. 1970;40:94–9.
106. Dyer JE, Reed JH. Alkali burns from illicit manufacture of GHB [abstr]. J Toxicol Clin Toxicol. 1997;35:553.
107. Galloway GP, Frederick SL, Staggers J. Physical dependence on sodium oxybate [letter]. Lancet. 1993;343:57.
108. Dyer JE, Andrews KM. Gamma-hydroxybutyrate withdrawal [abstr]. J Toxicol Clin Toxicol. 1997;35:553.
109. Dyer JE, Roth B, Hyma BA. GHB withdrawal syndrome: eight cases [abstr]. J Toxicol Clin Toxicol. 1999;37:650.
110. Craig K, Gomez HF, McManus JL, Bania TC. Severe gamma-hydroxybutyrate withdrawal: a case report and literature review. J Emerg Med. 2000;18:65–70.
111. Friedman J, Westlake R, Furman M. "Grievous bodily harm": gammaq-hydroxybutyrate abuse leading to a Wernicke-Korsakoff syndrome. Neurology. 1996;46:469–71.
112. Li J, Stokes SA, Woeckener A. A tale of novel intoxication: a review of the effects of gamma-hydroxybutyric acid with recommendations for management. Ann Emerg Med. 1998;31:729–36.
113. Harraway T, Stephenson L. Gamma-hydroxybutyrate intoxication: the gold coast experience. Emerg Med. 1999;11:45–8.
114. Badcock NR, Zotti R. Rapid screening test for gamma-hydroxybutyric acid (GHB, Fantasy) in urine [letter]. Ther Drug Monit. 1999;21:376.
115. Zvosec DL, Smith SW, McCutcheon JR, Spillane J, Hall BJ, Peacock EA. Adverse events, including death, associated with the use of 1,4-butanediol. N Engl J Med. 2001;344:87–94.
116. Marwick C. Coma-inducing drug GHB may be reclassified. JAMA. 1997;277:1505–6.
117. Dyer JE. Gamma-hydroxybutyrate (GHB). In: Olson KR, editor. Poisoning and drug overdose. Stamford, CT: Appleton & Lange; 1999. pp. 179–81.
118. Henderson RS, Holmes CM. Reversal of the anaesthetic action of sodium gamma-hydroxybutyrate. Anaesth Intensive Care. 1976;4:351–4.
119. Lelkens JP. A simple, cheap, effective and safe procedure for general anesthesia. Acta Anaesthesiol Belg. 1976;27:25–34.
120. Holmes CM, Henderson RS. The elimination of pollution by a noninhalational technique. Anaesth Intensive Care. 1978;6:120–4.
121. Wood DM, Dargan PI, Button J, Holt DW, Ovaska H, Ramsey J, et al. Collapse, reported seizure – and an unexpected pill. Lancet. 2007;369:1411–3.
122. White R, Standen O. Piperazine in the treatment of threadworms in children; report on a clinical trial. Br Med J. 1953;2:755–8.
123. Standen O. Activity of piperazine, in vitro, against Ascaris lumbricoides. Br Med J. 1955;2:20–2.

124. Campbell H, Cline W, Evans M, Lloyd J, Peck A. Comparison of the effects of dexamphetamine and 1-benzylpiperazine in former addicts. Eur J Clin Pharmacol. 1973;6:170–6.
125. Gee P, Fountain J. Party on BZP party pills in New Zealand. N Z Med J. 2007;120:u2422.
126. Alansari M, Hamilton D. Nephrotoxicity of BZP-based herbal party pills: a New Zealand case report. N Z Med J. 2006;119:U1959.
127. Baumann M, Clark R, Budzynski A, Partilla J, Blough B, Rothman R. Effects of "Legal X" piperazine analogs on dopamine and serotonin release in rat brain. Ann N Y Acad Sci. 2006;1025:189–97.
128. Tekes K, Tóthfalusi L, Malomvölgyi B, Hermán F, Magyar K. Studies on the biochemical mode of action of EGYT-475, a new antidepressant. Pol J Pharmacol Pharm. 2007;30:203–11.
129. Lyon R, Titeler M, McKenney J, Magee P, Glennon R. Synthesis and evaluation of phenyl- and benzylpiperazines as potential serotonergic agents. J Med Chem. 1986;29:630–4.
130. Brennan KA, Lake B, Hely L, H. S., Jones K, K., Gittings D, Colussi-Mas J, et al. N-Benzylpiperazine has characteristics of a drug of abuse. Behav Pharmacol. 2007;18:785–90.
131. Brennan K, Johnstone A, Fitzmaurice P, Lea R, Schenk S. Chronic benzylpiperazine (BZP) exposure produces behavioral sensitization and cross-sensitization to methamphetamine (MA). Drug Alcohol Depend. 2007;88:204–13.
132. Preissel U, Preissel HG. Brugmansia and Datura: Angel's Trumpets and Thorn Apples. Buffalo, NY: Firefly Books; 2002. pp. 106–129.
133. van Wyk BE. Handbuch der Arzneipflanzen. Stuttgart: Wissnschaftliche Verlagsgesllschaft; 2004.
134. Barker SA, Monti JA, Christian ST. N,N-Dimethyltryptamine: an endogenous hallucinogen. Rev Neurobiol. 1981;22:83–110.
135. Biocca EF, Galeffi C, Montalvo EG, Marini-Bettòlo GB. Sulla sostanze allucinogene impiegata in Amazzonia. Nota 1. Osservazioni sul paricà dei Tukfino e TariSna del bacino del Rio Uaupds. Annali di Chimica. 1964;54:1175–8.
136. Ott J. Pharmahuasca: human pharmacology of oral DMT plus harmine. J Psychoact Drugs. 1999;31:171–7.
137. Callaway JC, Grob CS. Ayahuasca preparations and serotonin reuptake inhibitors: a potential combination for adverse interaction. J Psychoact Drugs. 1998;30:367–9.
138. Passie T, Seifert J, Schneider U, Emrich HM. The pharmacology of psilocybin. Addict Biol. 2002;7:357–64.
139. Fischer R, England SM, Archer RC, Dean RK. Psilocybin reactivity and time contraction as measured by psychomotor performance. Drug Res/Arzneimittelforsch. 1966;16:180–5.
140. Wittmann M, Carter O, Hasler F, et al. Effects of psilocybin on time perception and temporal control of behaviour in humans. J Psychopharmacol. 2007;21:50–64.
141. Wackermann J, Wittmann M, Hasler F, Vollenweider FX. Effects of varied doses of psilocybin on time interval reproduction in human subjects. Nuerosci Lett. 2008;435:51–5.
142. Griffiths R, Richards W, Johnson M, McCann U, Jesse R. Mystical-type experiences occasioned by psilocybin mediate the attribution of personal meaning and spiritual significance 14 months later. J Psychopharmacol. 2008;22:621–32.
143. Alper KR, Lotsof HS, Kaplan CD. The ibogaine medical subculture. J Ethnopharmacology. 2008;115:9–14.
144. Popik P, Skolnick P. Pharmacology of ibogaine and ibogaine-related alkaloids. In: Cordell GA, editor. The alkaloids. San Diego, CA: Academic; 1998. pp. 179–231.
145. Alper KR. Ibogaine: A review. In: Alper KR, editor. The alkaloids. San Diego, CA: Academic; 2001. pp. 1–38.
146. Alper KR, Lotsof HS, Frenken GMN, Luciano DJ, Bastiaans J. Treatment of acute opioid withdrawal with ibogaine. Am J Drug Addict. 1999;8:234–42.
147. Mash DC, Kovera CA, Pablo J, Tyndale RF, Erwin FD, Williams IC, et al. Ibogaine: complex pharmacokinetics, concerns for safety, and preliminary efficacy measures. Neurobiological Mechanisms of Drugs of Abuse 2000. pp. 394–401.
148. Glick SD, et al. (±)-18-Methoxycoronaridine: a novel iboga alkaloid congener having potential anti-addictive efficacy. CNS Drug Rev. 1999;5:27–42.

149. Helsley S, Fiorella D, Rabin RA, Winter JC. Behavioral and biochemical evidence for a nonessential 5-HT2A component of the ibogaine-induced discriminative stimulus. Pharmacol Biochem Behav. 1998;59:419–25.
150. Mach RH, Smith CR, Childers SR. Ibogaine possesses a selective affinity for sigma 2 receptors. Life Sci. 1995;57:PL57–62.
151. Maciulaitis R, Kontrimaviciute V, Bressolle F, Briedis V. Ibogaine, an anti-addictive drug: pharmacology and time to go further in development. A narrative review. Human Exp Toxicol. 2008;27:181–94.
152. Nichols DE. Hallucinogens. Pharmacol Ther. 2004;101:131–81.
153. Monte AP, Waldman SR, Marona-Lewicka D, et al. Dihydrobenzofuran analogues of hallucinogens. 4. Mescaline derivatives. J Med Chem. 1997;40:2997–3008.
154. Diaz J. How drugs influence behavior. Englewood Cliffs, NJ\Prentice Hall; 1996.
155. Perrine DM. Hallucinogens and obsessive-compulsive disorder. Am J Psychiatry. 1999;156:1123.
156. Buckholtz NS, Zhou DF, Freedman DX, Potter WZ. Lysergic acid diethylamide (LSD) administration selectively downregulates serotonin 2 receptors in rat brain. Neuropsychopharmacology. 1990;3:137–48.
157. Sewell RA, et al. Response of cluster headache to psilocybin and LSD. Neurology. 2006;66: 1920–2.

Part V
How to Demask the Patient with an Aberrant Drug Use

Once there is the suspicion of a possible abuse of a drug, one should gather information from multiple sources to validate the individual responses or concerns of others. Similarly as in opioid dependence, the same statements do apply for cocaine, or amphetamine addiction. An unanticipated positive urine drug screen or worsening results are often indications that drug abuse is occurring [16]. However, contrary to belief, no behavior is an absolute predictor of aberrant drug use or of an addictive disease. Therefore, a differential diagnosis must be made.

Stigma of Persons with a Potential Drug Addictive Behavior

Although not pathognomonic, the following behavior patterns could indicate an addiction:

1. The role nicotine plays in the patient's life should be established, including a question about the time of day when he or she has the first cigarette. The severity of tobacco addiction correlates with the time to first cigarette of the day, the most important cigarette of the day, and the number of cigarettes smoked daily. A person who smokes within minutes after arising, or even before getting out of bed, and whose first cigarette of the day is the most important are often severely addicted [2].
2. A positive family history of substance use, whether for alcohol, illicit drugs, as well as mental health and emotional problems are potential risk factors for an addictive disease in all populations [17, 18]. Therefore, it is necessary to ask the person about both alcohol and illicit drug abuse among family members (i.e., parents, siblings, children) and second-degree relatives (i.e., grandparents, uncles, aunts, cousins) [18].
3. The participation or a recommended participation in a drug abuse treatment programs should be determined. Any person who has undergone any detoxification program in the past is at higher risk for a relapse [19].

4. Since nearly all drugs of abuse alter sexual function [1], and although the perception exists that drugs enhance sexual performance, long drug exposure typically decreases performance resulting in impotence or sexual dysfunction.

The Alcohol Addictive Patient

Signs of Alcohol Co-abuse in Addiction

Alcohol consumption and abuse can show variety of cutaneous manifestations, including palmar erythema and spider angioma, also known as telangiectasis, or spider nevus [20]. Aside, there are many endocrine changes in patients with chronic alcoholism with hypogonadism and/or hyperestrogenism as seen in male patients. While hypogonadism manifests in loss of libido/potency, testicular atrophy, reduced fertility, and reduced facial hair growth [20], signs of hyperestrogenism are characterized by gynecomastia, vascular spiders, changes in fat distribution, loss of body hair, and change of pubic hair to a female distribution [20]. This is in contrast to female patients with alcoholism, who rarely demonstrate signs of masculinization. They, however, may demonstrate breast atrophy or menstrual irregularities [20]. In both patients with alcoholic hepatitis, 30% demonstrate an enlarged, smooth occasionally tender liver [2].

And since a drug history is often combined with alcohol abuse, an initial screening should be included. For example, by using the CAGE questionnaire, the likelihood for alcohol abuse problems can be assessed [16, 21, 22]. The **CAGE** questionnaire consists should of simple questions related to quantity and frequency of patients who drink alcohol [21, 23]:

1. Have you felt the need to cut (C) down on your drinking (or drug use)?
2. Have people annoyed (A) you by criticizing your drinking (or drug use)?
3. Have you ever felt bad or guilty (G) about your drinking (or drug use)?
4. Have you ever needed an eye-opener (E) the first thing in the morning to steady your nerves or get rid of a hangover?

If two of four questions are positive, the diagnosis of a history of alcohol abuse or dependency shows a sensitivity of 74% and a specificity of 91%.

Another screening tool to assess the potential for substance abuse is the **Trauma Test**. By using a noninvasive method for obtaining information it can provide an important adjunct to diagnosis. Because of its brief structure it provides a cost-effective screening procedure in clinical practice composed of the following five questions:

Since your eighteenth birthday, have you:

1. Had any fractures or dislocations to your bones or joints (excluding sports injuries)?
2. Been injured in a traffic accident?
3. Injured your head (excluding sports injuries)?

4. Been in a fight or assaulted while being intoxicated?
5. Been injured while being intoxicated?

A positive response to two or more questions indicates a strong potential for alcohol (or drug) abuse. Patients with a history of driving under the influence or a history of two or more non-sport-related traumatic events (after age 18 years) are considered a high-risk population for substance abuse [16]. Clinicians should use this tool in conjunction with laboratory tests and a brief questionnaire that directly inquires about problems related to alcohol use [16, 22].

Other collateral information obtained from family, employers, and previous medical records [16], such as sudden loss of a job, frequent job changes for no apparent reason, and unexplained financial or family problems are often related to substance abuse [16]. In addition, if an eating disorder, together with sexual abuse are present, one should look for the third, a substance abuse [24]. It is because of this combination that researchers have recommended that all women who enter a substance abuse treatment should be screened for eating disorders [25].

Another instrument for screening of substance use is the CRAFFT test. It composes of six-questions of the "yes" or "no" type. Like the CAGE, the **CRAFFT** test takes about 1 min to complete, and can be incorporated into any type of evaluation [26]. While items 1, 2, and 5 pertain to personal drinking or drug intake, and are heavily backed by adolescents with substance abuse only item 6 relates directly to the criteria for abuse [26]. Being an appropriate screening tool for adolescents [27], it is verbally administered, simple to score, easy to remember and is composed of the following six questions:

1. Have you ever been in a car (C) driven by someone (including yourself) who was "high" or had been using alcohol or drugs?
2. Do you ever use alcohol or drugs to relax (R), to feel better about yourself, or to socialize?
3. Do you ever use alcohol or drugs while you are alone (A)?
4. Have you ever forgotten (F) things you did while being under the influence of alcohol or drugs?
5. Does your family or friends (F) tell you that you should cut down on your drinking or drug use?
6. Have you ever gotten into trouble (T) while using alcohol or drugs?

A CRAFFT score of two or higher identifies problems of abuse, or dependence with a sensitivity of 76% and a specificity of 94%. If a screen is positive it should be followed by a more complete substance use history [27].

Although an optimal marker of excessive alcohol consumption has not been found, γGT (gamma glutamate transferase) is the most widely used. Elevated levels, however, can also be seen in nonalcoholic liver disease, hepatobiliary disorders, obesity, diabetes, hypertriglyceridemia, and/or the use of liver microsome-inducing drugs [28]. Another frequently used marker for alcohol abuse is an elevated mean corpuscular volume (MCV). Because the combinations of more than one marker give a better sensitivity both, MCV and γGT greatly improve the potential to identify the person with excessive alcohol consumption. It also permits to identify false-negative subjects with low γGT levels who, however, have high alcohol consumption [29].

Secondary markers related to alcoholic hepatitis, show an activity of the serum alanine aminotransferase (ALT), which is depressed in relation to aspartate aminotransferase (AST). Thus, an AST/ALT ratio >2 may indicate alcoholic hepatitis [2]. Also, a jaundiced eye with a small pupil is suggestive of hepatitis C resulting form intravenous drug abuse (Fig. 4).

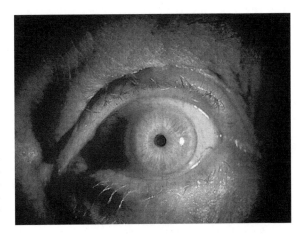

Fig. 4 A pin-point constricted pupil is characteristic for opioid use. The photograph was taken in normal light, so the pupil is not an artifact of flash photography (Adapted from [30])

Positive Signs and Symptoms of Hard Drug Abuse

Drug abusers frequently develop typical medical sequelae of their addiction. For instance, smoking or snorting cocaine and other drugs can cause respiratory problems, atrophy of the nasal mucosa, and perforation of the nasal septum [16]. On the other hand, needle marks may be present on the skin from recent injections, or "*tracks*" may be present over veins from past injections. However, injections are not always confined to obvious sites. Many users will inject into the axilla, under the tongue, under the breast, in the legs, and even into the dorsal vein of the penis [16]. In addition, many heroin addicts begin with subcutaneous injections called "*skin popping*" and return to this mode of application when extensive clogging makes their veins inaccessible (Fig. 8). As addicts get more desperate for the "*kick*", cutaneous ulcers may be found in improbable sites [2].

Signs and Symptoms for Abusive Drug Behavior

There are specific signs and symptoms to look for in order to detect a past or a present abusive behavior of drugs. A perforated septum (Fig. 5) or even continual sniffing and/or a permanent clogged up nose may be the result of piercing, trauma, or repeated snorting of cocaine, which is both vasoconstrictive and locally irritating to the nasal mucosa.

Also, due to necrosis and venous thrombosis after repeated use, the abuser chooses multiple injection sites. In an effort to conceal signs of drug abuse from healthcare professionals, family, or law enforcement, needle tracks (multiple, linear, and often hyperpigmented scars; Fig. 6a and b) are found at unusual places such as over the arms, wrists, axillae, neck, groin, between the toes, on the breasts, and even the dorsal vein of the penis.

Because the components of a pill do not dissolve well and when used for intravenous injection clog blood vessels, block blood flow (Fig. 7), with an irritation of the intimae of blood vessels resulting in vascular inflammation and abscess.

Since the typical progress of drug abuse is from snorting or smoking, to intravenous injection, this is followed by "*skin-popping*" when no access to veins remains.

Fig. 5 Picture of the left nares with the left turbinate with some ulcerations, and unexpectedly also the right turbinate. Due to excessive cocaine snorting the septum is perforated showing a healed hole (Adapted from [30])

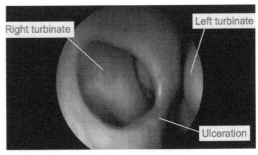

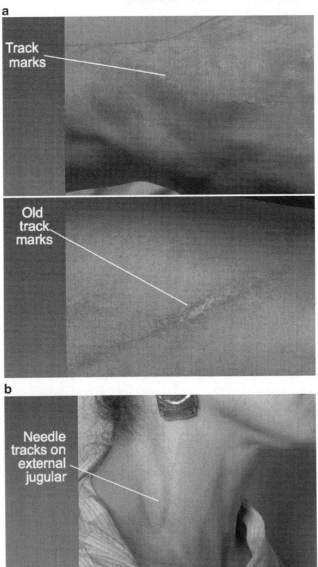

Fig. 6 (**a**) Fresh (*top*) and old track tracks (*bottom*) from intravenous drug use on the dorsum of the hand (Adapted from [30]) (**b**) Needle tracks at accessible veins are typical for the injection of illicit drugs (Adapted from [30])

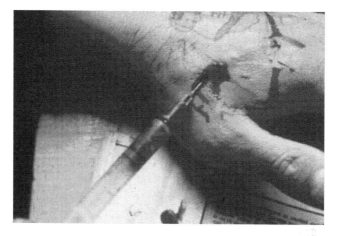

Fig. 7 The abuser trying to inject into a partly clogged-up vein

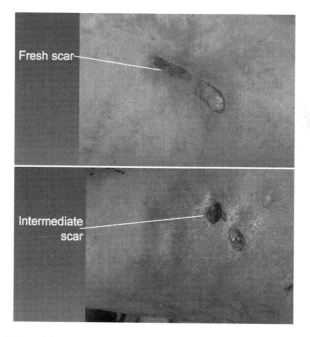

Fig. 8 Fresh (*top*) and intermediate *skin-popping* scar (*bottom*) resulting from subcutaneous injection of drugs (Adapted from [30])

As a result the abusers runs out of accessible veins, resorting to "skin popping", i.e. the subcutaneous injection of drugs. *Skin-popping* scars are irregular or round and look like small-pox vaccination scars (Fig. 8), except they are multiple and are not found

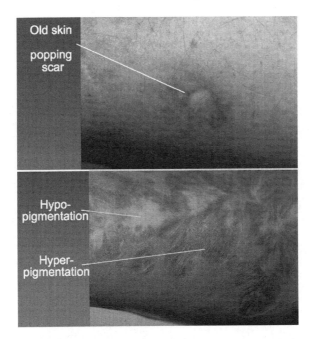

Fig. 9 Old *skin-popping* scars with hypo- and/or hyperpigmentation resulting from subcutaneous injection of drugs (Adapted from [30])

where expected over the deltoid muscle, and can be found all over the body being a common cause of abscesses.

Also, an old *skin-popping* scar on a light skinned person, can be mistaken for a cigarette burn, but one would normally expect to find a single scar instead of multiples flowing together (Fig. 9).

Injecting drug users are also at risk for infective endocarditis and valvular murmurs [2]. Since commonly alcohol is being used with illegal drugs this often leads to inadequate nutrition. This in turn is followed by vitamine deficiency and the drug addict presents a wasting syndrome [2].

Urine Drug Screening (UDS): Identifying the Person with Illicit Use

Prerequisites for Use of UDS

Urine drug screening is designed to detect illicit and/or licit non-prescribed drug abuse, and not to monitor adherence to treatment regimens. Actually the term *"urine drug screening"* is a misnomer since it implies the screening for all drugs [31]. In reality, however, it is *not possible* to prove the presence or absence of all drugs as the testing procedure is unlimited and is still growing [32].

All urine drug testing is not equal, and there is no *"standard urine drug test (UDT)"* that is suitable for all purposes and settings. Instead, a multitude of options exist that can be adapted to their clinical needs, while standard tests only exist in federal regulated industries [33]. Therefore, one must indicate to the testing laboratory whether the presence of any particular substance or group of substances is suspected or expected [32]. Thus, strong lines of communication with the laboratory personnel or the manufacturer of testing devices are essential to learn what can and what cannot be expected of a particular test.

The purpose of urine drug testing in clinical practice, where the majority of persons are going to tamper with their urine sample, is to enhance care. However, certain things have to be observed in order to improve the reliability of results:

1. Random collection is preferred, so a person is not told in advance of the request for the urine sample.
2. Unobserved urine collection is usually acceptable, but observed collection, especially in high-risk patients, may be necessary.
3. In order to reduce the potential for specimen dilution, the collection facility should not contain a basin with running water, and a blue pigment should be added to the toilet water.
4. An unusually hot or cold specimen, a small sample volume, or unusual color should raise suspicion. Although urine specimens will cool down to room temperature, the temperature of an urine sample should only fall within the range of 90–100°F within 4 min. A temperature strip built into the urine collection container can check this.

5. Urinary pH should be within the range of 4.5–8.0, and urinary creatinine concentration should be greater than 20 mg/dL or less than 20 mg/dL It can be considered as diluted it is <5 mg/dL not being consistent with human urine [34]. Urinary creatinine measurement can be done by an automated, inexpensive, and well-characterized method to test specimen validity.
6. The color of a urine specimen is related to the concentration of its constituents. Urine may be colored as a result of endogenous/exogenous substances derived from food pigments, medications, or disease states that produce excessive substances. It can be colorless as a result of excess hydration of a diet, a medical condition, or of water intake. In order to prevent diluted urine samples should be taken in the early morning. Any results outside of these ranges should be discussed with the patient and/or laboratory, as necessary [34].
7. The unexpected negative urine test. There are multiple reasons for a negative urine drug screen, including the cut-off points (see later) used by the testing laboratory. In addition, the specificity and sensitivity of immunoassays vary considerably depending on the assay type and the specific test being performed. Because antibodies as being used in an urine test are seldom specific to a single drug or drug metabolite, any cross-reactions may cause false positive results.
8. Any positive result, which is based on immunoassay test alone is referred to as *"presumptive positive"* indicating that it must be confirmed by a different assay technique such as gas chromatography and/or mass spectrometry (GC/MS) with different sensitivity, specificity, and reliability [33]. For instance, a positive opiate urine screen cannot distinguish between morphine, codeine, or heroin; this has to be done using the GC/MS technique [33].
9. Although GC/MS can quantitate the level of drug in the urine, no conclusions can be drawn as to how much was consumed since the levels of drugs or drug metabolites in the urine are affected by numerous factors.

Types of Urine Drug Testing

Urine drug screening is typically a two-step procedure. The first step uses commercially available immunoassay to detect the presence of a drug or metabolite in the urine. Once specimens are identified as positive they have to be confirmed using gas chromatography/mass spectrometry [33]. The first stage analysis is known as a screening method performed by the "*Enzyme Linked Immuno Sorbant Assay*" (ELISA) that detects not only traces of drugs but also their metabolites. It is important to note, that the time-span where drugs can be detected as positive varies in regard to the specimen being used for screening (Fig. 10).

The specificity and sensitivity of immunoassays vary depending on the type of assay and on the specific test being performed. The primary disadvantage of immunoassays is that the antibodies are seldom specific to one single drug or one drug metabolite. Therefore, the antibodies may bind with other substances. It is because of an immunoassay based positive result the term "*presumptive positive*" is being used since other factors such as cross-reactivity and difference in sensitivity and specificity among immunoassays does exist. Such results must be confirmed by a more specific method.

Commercially available immunoassays, such as EMIT® II, KIMS®, CEDIA®, DRI®, and AxSYM®, use antibodies to detect the presence of a drug or metabolite in urine [33] with the principal advantage of simultaneously and rapidly test for drugs in urine specimen. The principal disadvantage, however, is that they vary in the range of compounds detected among each other, some detecting specific drugs, while others recognize only classes of drugs. It is important to know which immunoassay is being used because sensitivity and specificity varies markedly among the different testing devices. Because the ability of an immunoassay to detect a drug depends on the drug's concentration in the urine, any response above the cut-off is considered positive and any response below the cut-off is negative [35, 36]. For example, if the cut-off in an immunoassay is set at 50 ng/mL, a drug concentration of 49 ng/mL may be indicated as being negative [34]. In addition, immunoassays also show cross-reactivity; i.e., substances with similar, and sometimes different, chemical structure may cause a test to incorrectly appear positive for the target drug. Also, it is important to note that immunoassays are not static, they are continually updated by the manufacturer and may even be altered by the testing laboratory [37].

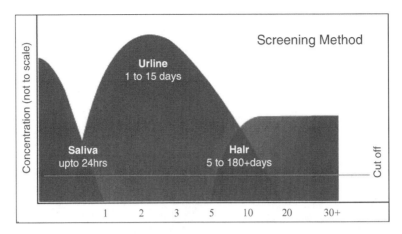

Fig. 10 The time scale (in days) put in relation to the cut-off values of different specimens for positive testing of cocaine abuse

If a positive result appears in the immunoassay than a second stage process known as a *"confirmation test"* is performed using either gas chromatography or mass spectrometry (GC/MS). This second test has a cut-off level that eliminates specimens containing drug levels from environmental contamination [33]. In order to be considered positive, the urine sample must show the presence of the drug in these two different assays. This two-step procedure protects against false positive reports.
Although quantitative results are provided by *GC/MS*, there is no correlation between urine drug concentration and dose taken [32, 38]. This is because, that aside from daily dose intake, there are many factors that determine urinary concentration of drugs and their metabolites. However, even in gas chromatography/mass

> *Gas chromatography or mass spectrometry (GC/MS) by Forensic Science Service is considered the gold standard for testing a urine sample. It is usually the only test, which is accepted in court. GC/MS is the tool of choice because it has good sensitivity, selectivity, specificity, a high degree of standardization, high sample throughput, and a stable instrument performance.*

spectrometry (GC/MS) variation in drug concentration can exist. This is because of the difference in methods being used for performing assays, the variation in factors affecting an assay, the potential for carryover from another specimen, and the difference in cut-off concentration chosen by the laboratory [33]. For example, most laboratories have a limit of detection below which they will not report any drug as being positive [31]. And although laboratory-based specific drug identification with GC/MS confirms the presence of a drug and/or its metabolite(s), it does not necessarily identify another drug of the same class. For example, if morphine is the

Table 2 Open window for detecting different abused drugs in the urine (PCP = phencyclidine; TCA = tricyclic antidepressants; EDDP = 2-ethylidine-1,5-dimethyl-3,3-diphenyl-pyrrolidine)

Amphetamine	1–3 days
Barbiturates	4–8 days; in the case of chronic abuse several weeks
Benzodiazepines	3 days after therapeutic dosage; up to 4–6 weeks in case of long-term use
Buprenorphine	2–6 days
Cannabis	The half-life is several weeks. Due to its good fat solubility, strong usage of cannabis can be positively detected in urine tests even after 20–30 days
EDDP (methadone metabolite)	2–7 days
Cocaine	Half-life ca. 90 min. With a cut-off of 300 ng/ml, detectable after 2–4 days
Methadone	2–5 days
Opiates/opioids	2–3 days
PCP	2–3 days
TCA	2–3 days

opiate causing a positive immunoassay, other drugs are not included in a given assay; i.e., oxycodone, hydromorphone, hydrocodone, and fentanyl are not identified [33].

Half-Life of Detection and Cut-Offs in Urine Drug Screening

Immunoassay drug screening tests offer a fast and reliable method to demask a possible drug consumer. Various commercially tests are available with up to 12 different drugs and medications as well as different combinations. Depending on the time of abuse there is a large variability in half-life during which the agent can be detected in urine specimen (Table 2).

The cut-offs presented in the next table are those used for immunoassays and which are mandated by NIDA (National Institute of Drug Abuse) for urine drug testing programs [33]. The detection time of a drug in urine indicates how long after administration a person excretes the drug and/or its metabolite(s) at a concentration above cut-off concentration in a specific test [35, 39]. Although this is governed by several factors, including dose, route of administration, metabolism, urine volume, and pH, the detection time of most drugs in the urine is less than 5 days, typically 1–3 days [11, 35, 36]. Long-term use of lipid-soluble drugs, such as marijuana, diazepam, or phencyclidine, may enlarge the window of detection to as long as a month [11, 33, 35]. For instance, a cut-off value of 1,000 ng/mL in urine samples can be positive for amphetamine for up to 5 days after intake [35]. On the average, following smoking of only one marijuana cigarette, the active ingredient tetrahydrocannabinol (THC) may be detectable for 2–4 days in the urine. However, more frequent users may stay positive for up to a month [35]. Street doses of cocaine may be detectable for up to a week [35], and the detection time for opiates is about 1–2 days [35, 40]. It should be noted, that a positive "*opiate*" result reveals the presence of codeine and/or

Table 3 Cut-off values of different agents used in urine drug screening (EDDP = 2-ethylidine-1,5-dimethyl -3,3-diphenyl-pyrrolidine; AMP = amphetamine; MET = methamphetamine: TCA = tricyclic antidepressants)

Drug	Cut-off value (ng/ml)
AMP300	300
Amphetamines	1,000
Barbiturates	300
Benzodiazepines	300
Buprenorphine	20
Cannabinoides	50
EDDP	100
MET300	300
Methadone	300
Methamphetamines	1,000
Cocaine	300
Opiates	300
TCA	1,000

morphine, which is a metabolite of heroin. Duration of detectability of phencyclidine is 8 days, though in chronic users it may be detectable for up to a month [11, 33]. According to the individual cut-off values of drugs, they can be detected in urine down to a specific concentration. Anything below this specific cut-off value will not be identified (Table 3).

Interpretation of Positive Urine Drug Tests (UDT)

Based on the predetermined cut-off concentrations a qualitative immunoassay drug panel indicates each sample as being either positive or negative for a particular drug or a drug class [11, 35, 36]. A positive UDT reflects recent use of the drug, because most substances in urine have detection times of only 1–3 days. Positive results, however, do not provide enough information to determine the exposure time, the consumed dose, or the frequency of abuse [35]. Ideally, a UDT would be positive if the patient took the drug (true positive) and negative if the drug was not taken (true negative). However, false-positive or false-negative results can occur for a number of reasons. It therefore is important to interpret the UDT results carefully [11].

Considerations When Having a Positive Urine Drug Test

Controversies exist regarding the clinical value of UDTs, partly because most current methods are designed for, or adapted from, forensic or workplace deterrent-based testing for illicit drug use. These are not necessarily optimized for widespread clinical applications [31]. However, when used with an appropriate level of understanding, and with follow-up documentation, UDS can improve the healthcare professionals' ability to diagnose substance misuse, abuse, or addiction. The following points should be considered when having a positive result:

1. Ingestion of poppy seeds in cakes or cookies result in a positive opiate screening test. This however, has been addressed by a higher cut-off change since 1998.
2. Codeine is metabolized to morphine, so both substances may occur in urine following codeine use [11, 33, 38]. Therefore a prescription for codeine may explain the presence of both codeine and morphine in the urine.
3. In contrast, a prescription for codeine does not usually explain the presence of only morphine (although samples collected 2–3 days after codeine ingestion may contain only morphine). Morphine alone is most consistent with ingestion of morphine or heroin.
4. The detection of codeine alone is possible because there is a small proportion of patients that lack the cytochrome P450 2D6 enzyme necessary to convert the prodrug codeine to its active metabolite morphine.
5. A prescription for morphine *cannot* account for the presence of codeine. And although codeine is metabolized to morphine, the reverse does *not* occur.
6. However, prescribed codeine may explain the presence of codeine with trace amounts of hydrocodone, which can be produced as a minor metabolite of codeine [41].
7. Cocaine is a topical anesthetic clinically used in certain trauma, dental, ophthalmoscopic, and otolaryngologic procedures. A patient's urine may test positive for the cocaine metabolite benzoylecgonine after such a procedure for up to 2–3 days. However, a licensed healthcare professional must order its use, which can be checked through medical records or by contacting the treating healthcare professional. Since there is no structural similarity between other local anesthetics (e.g. procaine, lidocaine) and cocaine or benzoylecgonine, cross-reaction does not occur, and being exposed to a local anesthetic is **not** an explanation for the presence of benzoylecgonine in the urine.
8. There have been cases of cocaine ingestion by drinking tea made from coca leaves. Although such tea may be available for purchase, the product containing cocaine and/or its related metabolite(s) is illicit under US federal statutes and regulations, and so is not a valid explanation. Therefore persons are advised not to ingest coca tea.
9. Clinical interpretation of amphetamine and methamphetamine positive results can be challenging because of prescription use and structural similarities of many prescription and over-the-counter (OTC) products, including certain drugs used in the treatment of Parkinson's disease, diet agents (particularly Mexican diet pills), and decongestants [33, 38]. Knowledge of potential sources of amphetamine or methamphetamine can prevent misinterpretation of results. Therapeutic uses for amphetamine and methamphetamine include attention deficit disorder, treatment of exogenous obesity, and treatment of narcolepsy and CNS disorders [33]. Examples of prescription medications that contain amphetamine or methamphetamine are Adderall®, Benzedrine®, Dexedrine®, and Desoxyn®. In addition, immunoassay screening tests cross-react with various amphetamine-related

drugs that are not misused, such as dopamine, isoxsuprine, and ephedrine (an asthma medication). For example, dopamine, isoxsuprine (a vasodilator agent), or ephedrine, being used as an asthma medication. Especially the Vicks® Vapor Inhaler contains desoxyephedrine, which is the *l*-form of methamphetamine. Only by separation of the *d*- and the *l*-methamphetamine isomers, a 100% *l*-methamphetamine should be revealed when using Vicks® Vapor Inhaler. If however, there is >20% of *d*-methamphetamine it suggests an outside source of *d*-methamphetamine other than the inhaler [33, 42]. In any course the GC/MS confirmation distinguishes any cross-reacting compounds, assuring that results are not false positive.

10. Other amphetamine-like drugs that are sometimes misused and also cross-react are phenmetrazine, phentermine, fenfluramine, and mephentermine [33]. There are also several substances that are known to metabolize to amphetamine or methamphetamine including selegiline (for Parkinson's disease), benzphetamine, clobenzorex, dimethylamphetamine, fenproporex, and mefenorex [33] (Fig. 11).

11. Methamphetamine and amphetamine exist as two isomers, which are designated the *d*-form and the *l*-form. The *d*-form has a strong stimulant effect on the central nervous system and high abuse potential, while the *l*-form, when used in therapeutic doses, has a primarily peripheral action and is found in many OTC preparations

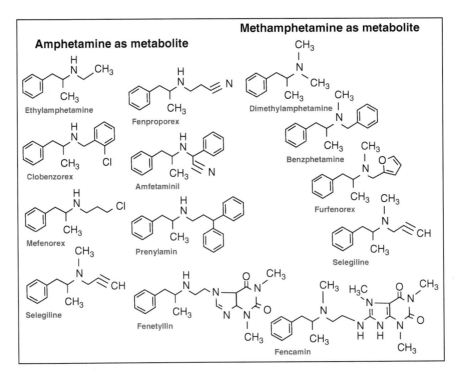

Fig. 11 Several compounds being taken as appetite suppressants, which readily metabolize to either methamphetamine or amphetamine

[33]. Routine testing, such as immunoassays or gas chromatography/mass spectrometry (GC/MS), does not differentiate between the *d*-form and the *l*-form of methamphetamine/amphetamine. For example, the OTC Vicks® Inhaler marketed in the US contains desoxyephedrine, the *l*-form of methamphetamine [33]. Persons undergoing urine drug testing therefore are advised not to use the Vicks® Inhaler or similar OTC products containing methamphetamine [34].

12. Only specialized tests, such as stereospecific chromatography, can distinguish between the two methamphetamine and amphetamine form [33]. The separation of the *d*- and *l*-isomers should reveal nearly 100% *l*-methamphetamine following Vicks® Inhaler use. If however, in a laboratory quantitative report there is >20% of *d*-methamphetamine present, it is suggestive of methamphetamine other than the inhaler. Illicitly manufactured methamphetamine/amphetamine is a mixture of *d*- and *l*-isomers. However, misuse of even the *l*-form of methamphetamine can have significant central activity and should be addressed [34].
13. Several years ago, the nonsteroidal anti-inflammatory drug ibuprofen was found to interfere with the EMIT® immunoassay test to cause false-positive results for marijuana. However, the problem has been corrected in the currently used EMIT® II, and ibuprofen no longer causes false positives in initial screening assays [33].
14. There have been reports of false-positive urine immunoassay test results for THC in patients receiving proton pump inhibitors, such as pantoprazole [43]. However, a confirmatory test such as GC/MS will not verify the positive immunoassay result.
15. Tetrahydrocannabinol (THC), the main ingredient of marijuana, has been prepared synthetically and marketed under the trade name Marinol® for the control of nausea and vomiting in cancer patients receiving chemotherapy and as an appetite stimulant for AIDS patients [44]. More specific testing would be required to distinguish between natural and synthetic THC. Also, passive smoke inhalation does not explain positive marijuana results at typical cutoffs at 50 ng/mL [33, 36]. Repeated positive results for marijuana should be viewed as evidence of ongoing substance misuse that requires further evaluation and possible treatment [34].
16. Although legally obtained hemp food products do not appear to be psychoactive, there have been concerns that ingestion of these food products, which contain traces of THC, may cause a positive UDT result for marijuana [33, 45, 46]. However, multiple studies have found that the THC concentrations typical in hemp seed products are sufficiently low to prevent a positive immunoassay result [45, 46]. Therefore, consumption of hemp food products generally is *not* a valid explanation for a urine immunoassay screen positive for marijuana.
17. False-positive results can also be reported because of technician or clerical error. These results may also occur because of cross-reactivity with other compounds found in the urine, which may or may not be structurally related; for example, some quinolone antibiotics, such as levofloxacin and ofloxacin, can potentially cause false-positive results for opiates in common immunoassays, despite no obvious structural similarity to morphine or codeine [33, 47, 48]. Fortunately,

identifying specific drugs or metabolites by gas chromatography/mass spectrometry (GC/MS) is not influenced by cross-reacting compounds.

Considerations When Having a Negative Urine Drug Test

A false-negative result is defined as a negative finding in a sample known to contain the drug of interest [34]. There are several possibilities, which have to be taken into consideration:

1. A negative urine test may be due to technician or clerical error or is the result of tampering with the urine sample. Methods being used by individuals attempting to influence the result of an UDT include adulteration and substitution of urine. Such manner should be suspected if the characteristics of the urine sample are inconsistent with normal human urine. Urine creatinine measurement is an inexpensive and well-characterized method to test specimen validity. In addition, pH, temperature, and use of an adulteration panel are also helpful to raise suspicion [11, 33].
2. If in a second outcome, the immunoassay is negative for opiates, and subsequent gas chromatography/mass spectrometry (GC/MS) fails to detect morphine, the negative result suggests that the person has not recently ingested morphine and may be reliable for not diverting or trafficking of drugs [34].
3. In a third outcome, the immunoassay is negative for opiates, but subsequent gas chromatography/mass spectrometry (GC/MS) is positive only for meperidine (pethidine). Such a result is suggestive that the patient is doctor shopping and/or misusing drugs, and requires further investigation to determine whether the patient presents a genuine case of substance misuse [34].
4. If in a fourth outcome, the immunoassay is positive for opiates and cocaine and a subsequent gas chromatography/mass spectrometry (GC/MS) confirms the presence of morphine and cocaine, the result suggests that the patient is abusing illicit drugs and may be misusing morphine or abusing heroin [34].

Type of Tests for Urine Drug Testing (UDT)

There are two types of test cards offered for UDT, both of which differ only in regard to how the specimen is being handled:

The Dip Test

In this test the user has to remove the cap from the test device and hold the absorbent tip into the urine for 10–15 s. There is no protective cap with every single test (Fig. 12).

The Drop Test

By using a pipette three drops of urine are dropped into each well. (Fig. 12). In both instances the results can be read after 5–10 min.

How to Interpret the Results

Due to the immunoassay reaction of the antibody with the drug in the urine specimen a band becomes visible at the respective level (Fig. 13).

Different multi-test cards, either as a dip or a drop-test are available, which can be selected in different compositions and combinations for individual drug testing (Fig. 14) such as opiates/opioids, cocaine with a cut-off at 200 ng/ml (COC 200), benzodiazepines, methadone, terahydrocannabinol with a cut-off value of 25 ng/ml (THC 25) amphetamines, barbiturates, tricyclic antidepressants (TCA), buprenorphine (BUP), methamphetamine (e.g. ecstasy), amphetamine with a cut-off at 200 ng/ml (AMP 200), methamphetamine with a cut-off at 300 ng/ml (MET 300),

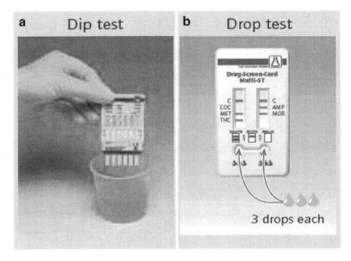

Fig. 12 Different Multi-Test cards using either the dip- or the drop-test method are available for drug analysis of various substances (nal von Minden® company, Moers/Germany)

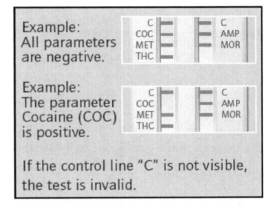

Fig. 13 Sample of a multiple-test card for determining various drugs (AMP = amphetamine; MOR = morphine; COC = cocaine; MET = methamphetamine; THC = tetrahydrocannabinol)

and/or the methadone metabolite EDDP. Only recently, nal vonMinden company has introduced an additional test for identification of the opioids tramadol and fentanyl.

Identification of the Methadone Metabolite in Substitution Therapy

During substitution therapy, EDDP (2-ethylidine-1,5-dimethyl-3,3-diphenyl-pyrrolidine; Fig. 15) is the main metabolite from methadone. Rather than identifying methadone, identification of its metabolite results in the following advantages:

Identification of the Methadone Metabolite in Substitution Therapy 263

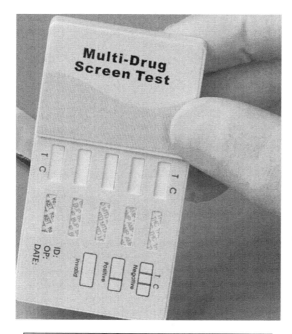

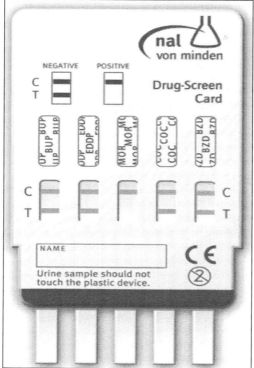

Fig. 14 (continued)

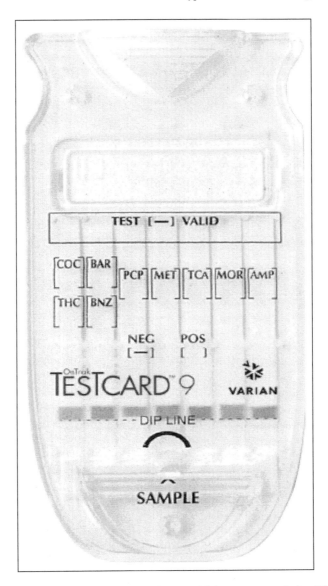

Fig. 14 Representative examples of three different multi-drug screen cards for different drugs: morphine (MOR), amphetamine (AMP), cocaine (COC), barbiturates (BAR), benzodiazepines (BENZ), methamphetamine (MET), morphine (MOR), phencyclidine (PCP), tetrahydrocannabinol (THC), tricyclic antidepressants (TCA), buprenorphine (BUP), and the methadone metabolite EDDP (By courtesy of MMC/The Netherlands, nal von Minden/Germany and Varian/USA)

1. During control, submission of foreign urine with methadone addition is no longer possible. By adding methadone to an otherwise negative urine sample, the methadone test will have a positive reading. The patient, while being in a

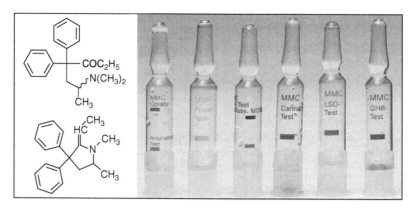

Fig. 15 Being part in a series of reagents for drug tests (MMC International Netherlands) methadone (*top*) and its metabolite EDDP (2-ethylidine-1,5-dimethyl-3,3-diphenylpyrrolidine (*bottom*), separately can be detected within a time range of 2–7 days after administration of the parent compound. Aside from GHB (=gamma hydroxybuturate), reagents for identification of LSD, opiates, ecstasy (MDMA), and amphetamine are also available

methadone program, he can claim to regularly take methadone. Because the EDDP test only reacts highly sensitive to the methadone metabolites, any attempt at manipulation remains negative. Even those individuals who metabolize quickly will be detected.

2. Some patients tend to metabolize methadone into EDDP very quickly (=fast metabolizer). In such a case, the methadone test remains negative, because there is no cross-reaction to the EDDP metabolite. With the EDDP test, however, these patients can be reliably identified.

Single Test Strips

In addition to *Multiple-Test Cards*, single test strips (i.e. sticks; Fig. 16) are also available for the qualitative detection of the following substances: Amphetamines, barbiturates, benzodiazepines, buprenorphine, cannabinoides, cocaine, EDDP, methadone, methamphetamines, opiates, PCP (phencyclidine), or TCA (tricyclic antidepressants). Having dipped the stick into the urine sample for 5–10 s, results can be viewed after 5–10 min.

- Contrary to other single test sticks, the Teststik® (Varian Inc/USA) does not use the colloidal gold technique, which requires timing and must be strictly followed or else test results will be inaccurate. The TesTstik® requires no mixing, clock-watching, or specialized operator training.

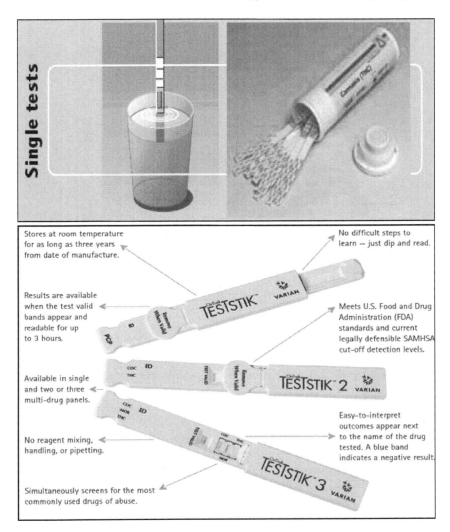

Fig. 16 Two examples of cost conscious single test sticks for qualitative analysis of a potentially abused substance in urine specimen

Test-Cup for Permanent Storage

Also, a TesTcup® II is available (Fig. 17) with an enhanced cup design is workable, as it features a leak proof lid and a one step process that make onsite testing easier, faster, and economical (Varian Inc/USA). Patent-pending, easy-to-read yellow stripes display test findings in less than 4 min and results can be read for up to 2 h. By scanning test results directly into a TesTrak® system, the drug test results can be securely stored permanently in a spreadsheet file, where data can be reviewed,

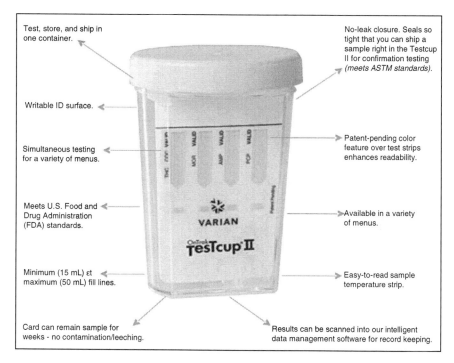

Fig. 17 The TesTcup®II allowing permanent storage of results for later comparison

analyzed, and charted for future retrieval. Test results can even be scanned and attached to an electronic image to the record for further verification.

The TesTcup®II from Varian Inc/USA, is available in different combinations for the following drugs: amphetamine, methamphetamine, benzodiazepines, cocaine, THC (marijuana), morphine, and PCP.

Detecting Manipulations of Urine Samples in UDT

The purpose of urine drug testing in clinical practice, where the majority is not going to tamper with their urine sample, is to enhance diagnostics or to identify the cause of an underlying emergency [34]. However, certain things can be done to improve the reliability of the results. Random collection is preferred, so the patient is not told in advance of the request for the urine sample. This can help to prevent the minority of patients, who might tamper with their sample from being prepared with adulterants or substituted specimens.

In all situations, some control of sample integrity is desirable, and it is imperative where the consequences of incorrect results have far-reaching implications [32]. Unobserved urine collection is usually acceptable, but observed collection may be necessary in certain situations or with high-risk patients. Ideally, the collection facility should not contain a basin with running water in order to reduce potential for specimen dilution, and blue pigment should be added to the toilet water. An unusually hot or cold specimen, small sample volume, or unusual color should raise concerns. Although urine specimens will cool to room temperature, the temperature of a urine sample within 4 min of voiding should fall within the range of 90–100°F, which can be checked at the time of collection by a temperature strip built into the urine collection container. Urinary pH should remain within the range of 4.5–8.0, and urinary creatinine concentration should be greater than 20 mg/dL A value less than 20 mg/dL is considered dilute and a value less than 5 mg/dL is not consistent with human urine [39]. Urinary creatinine measurement is an automated, inexpensive, and well-characterized method to test specimen validity. The color of a urine specimen is related to the concentration of its constituents. Urine may be colored as a result of endogenous/exogenous substances derived from food pigments, medications, or disease states that produce excessive analyses. It can appear colorless as a result of excess hydration because of diet, medical condition, or water intake. In the absence of underlying renal pathology, patients who repeatedly provide diluted urine samples should be advised to decrease water intake prior to testing and to provide samples in the early morning. Any results outside of these ranges should be discussed with the patient and/or laboratory, as necessary [34]. The test results indicate whether the urine sample has been chemically manipulated before undergoing the drug test. Thinning of the urine sample is most likely the most frequently used form of urine manipulation. Specific weight and creatinine are significant parameters in case of a suspicious in-vivo or in-vitro thinning of the urine sample (Table 4). Selective urine test strips are available, which can assess whether urine has been manipulated. The following parameters are being measured:

- pH value - Any adding of acids or basic solution will be identified.
- Nitrates - Nitrates can, if applied, falsify the test results.
- Specific weight - A thinning of the urine sample can be identified.
- Creatinine - Aside from in-vitro thinning, in-vivo thinning can also be recognized.
- Glutaraldehyde - This is found in various disinfectant solutions. It can falsify the test and will be detected with the strip. Bleach can affect the results if added to urine.
- Pyridine chlorochromate - A chemical that will falsify the test.

Table 4 Possible manipulations of urine and how they can be detected. Ultimately, manipulated urine cannot be used for analysis, as it causes false negative as well as false positive results

Type of manipulation	Substances, methods used	Characteristical changes
Thinning	Can occur in-vivo or in-vitro	Urine is very light Temperature is <32°C Specific weight less 1.01 g/ml Creatinine content less than 30 mg/dl
Addition of base	Bleaches like acetic acid or Domestos® are added to the sample	Decline in pH <8 A chloride scent may be present
Addition of acids	Acids like acetic acid or citric acid are added to the sample	Significant changes with an increase in pH >3
Addition of other substances	Soap or kitchen salt is added to the sample	Urine may have flakes Specific weight is greater than 1.035 g/ml

Analysis of Saliva, Hair and Sweat for Drug Testing

Urine is currently the most widely used and validated specimen for drug testing [11, 33, 38]. Although other technologies for use in different biologic specimens are marketed for drug testing, there is a lack of information on their false-positive and false-negative results, interferences, and cross-reactivity. Therefore, most of these techniques may not be appropriate for use in clinical practice [34].

Factors that affect the selection of a biologic specimen for drug analysis include ease of collection, analytical and testing considerations, and reliability of results [49, 50]. Advantages of saliva as a test sample include ease of collection, minimal personal invasiveness, collection under close supervision (prevention of tampering the sample), and limited pre-analytical handling (Table 5). However, drugs and their metabolites are retained for a shorter period and occur at lower concentrations when compared with urine [37, 49–51].

Hair Analysis for the Detection of Abused Drugs and Medications

Hair analysis provides another retrospective and long-term measure of drug use that is directly related to the length of hair [11, 50]. Hair analysis a suitable method for long-term detection of abuse because it has the advantage of being able to detect the abuse of drugs and/or medication over a long time period. Drugs and/or medications are stored in the hair follicles and grow along with the hair (Fig. 18). Since hair grows approximately 10–15 mm per month, past drug abuse can be determined by examining the respective segment of hair. For analysis, a pen-sized tuft of hair is required, which is cut directly at the scalp. Individual hairs are not sufficient for an analysis.

Hair analysis is most useful for the detection of rapidly eliminated drugs like heroin and cocaine, and where drug misuse is undetected by urine analysis [38]. However, darkly pigmented hair has a greater capacity to bind a drug than hair that is fair or gray. Therefore it is argued that hair analysis has a possible color

Table 5 Summary of pros ad cons for using saliva and and/or hair for drug testing

Saliva	Hair
Advantages	*Advantages*
– Ease of collection	– Long-term measure related to hair length
– Minimal invasive	– Analysis for weeks, months and even years
– Limited manipulation	
Disadvantages	*Disadvantages*
– Short period of preservation	– Dark hair greater capacity to bind drugs
– Fast degradation	– Irregular growth
	– Accessibility
	– Labor intensive sample preparation

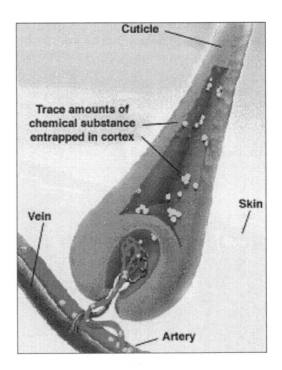

Fig. 18 Illustration of the human hair with trace amounts of an abused drug

bias [11, 50]. Other disadvantages of hair analysis include irregular growth, accessibility, and labor-intensive sample preparation. Gas chromatography is used for hair analysis using a mass-specific detector. To completion of drug hair analysis, a quick test and a confirmation analysis are available for different agents (Table 6).

Table 6 Excess of the above values is considered a positive result in hair analysis

Drug	Cut-off value (ng/mg hair)
Amphetamine/Methamphetamine/Ecstasy	1.10
Methadone	1.00
Opiate/Opioids	1.00
THC (Cannabis)	0.20
Benzoylecgonine (Cocaine)	0.50

Table 7 Summary of pros and cons for using sweat and/or blood for drug testing

Sweat	Blood
Advantage	*Advantage*
– Noninvasive	– Reduced of adulteration
– Cumulative over days to weeks	– Quantitative analysis possible
Disadvantage	*Disadvantage*
– Variation in sweat production	– Not open for rapid screening
– Risk of contamination	– Low concentrations are missed
	– Invasive procedure for collection

Analysis of Sweat for Abused Drugs and Medications

Sweat collection using a sweat patch provides another noninvasive, analysis of drug use over a period of days to weeks. It is mostly used for monitoring of drug use in addiction treatment or probation programs [11, 49]. The disadvantages of sweat analysis include varying sweat production and risk of accidentally removing or contaminating the collection device [49]. This is in contrast to blood samples where there is reduced likelihood that patients can influence test results. Also, blood analysis contains the possibility of an accurate determination of drug concentrations and the possibility to quantify the drug being abused [38]. However, blood samples are not always available for rapid screening (Table 7), because of elimination may show low drug concentrations, and requires an invasive technique [11, 38]. Therefore, blood analysis is not recommended for routine testing [11].

The relative detection times of drugs in biologic specimens are shown in the next figure. Blood and saliva maintain detectable levels of drugs for hours, urine for days, and sweat for weeks with a cumulative device, and hair and nails for several years (Fig. 19).

Sampling of saliva or sweat for analysis is not complicated. It is a painless, noninvasive procedure, often used as routine testing by the police and by custom officials. The result is indicated by a color change on the indicator strips, and can be read directly.

There are several systems for saliva testing on the market:

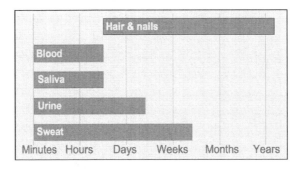

Fig. 19 Differences in detection times of abused drugs in various body specimen (Adapted from [49])

SmartClip® Test for Saliva Testing

The American company Honeywell provides a reliable saliva test where results are presented within in a short period of time. If a tested person has not taken drugs, proof of absence is (90% of all cases) is available within 60 s. Since hygienic requirements for drug testing, especially where sweat or saliva is collected are high, a hygienic concept that covers all aspects from unpacking of the single-use system to its disposal has evolved with development of the SmartClip®, EnviteC (Fig. 20). A number of features ensure a high degree of safety for both the tester and the tested person, since the SmartClip® combines the sampling and analysis in a single device.

The device has been tested over several years in cooperation with the police in their daily work and has proven to be practical. Its ease of use and high reliability make it the ideal device for preliminary drug testing. Safety and hygienic design is of particular importance in connection with acquiring and handling of the sample with special designed flaps that makes the SmartClip® hygienic (Fig. 21).

The innovative SmartClip® allows the testing of saliva as well as sweat samples for a variety of drugs. It also detects a number of substances on swab samples taken from people and surfaces of objects. The system is able to detects the following drugs:
- Cannabis
- Amphetamine
- Methamphetamine
- Ecstasy
- Cocaine
- Crack
- Opiates

There are two different SmartClip® tests available:

1. The SmartClip® Multidrug system for saliva, sweat and surface swab samples, which tests simultaneously for amphetamines, methamphetamines, ecstasy, cocaine, crack and opiates.

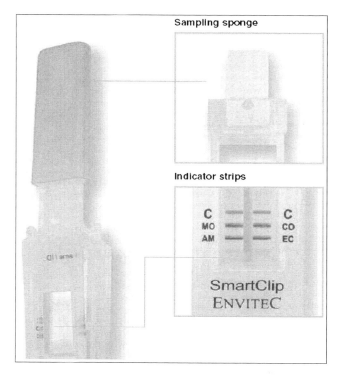

Fig. 20 The SmartClip® fort saliva or sweat testing, specifically developed to meet the day-to-day requirements of users

2. The SmartClip® THC/AM system for sweat and surface swab samples, which tests simultaneously for cannabis and amphetamines.
   - Both tests have the following cut-off values (Table 8).

Table 8 Different cut-off values of two SmartClip® tests for use in saliva and sweat **SmartClip® multidrug**

Substance	Saliva cut-off (ng/sample)	Sweat/swab sample cut-off (ng/sample)
Amphetamines	50	20
Methamphitamines	100	40
Cocaine	20	8
Morphine	40	16

SmartClip® THC/AM	
Substance	Sweat/swab sample cut-off (ng/sample)
Cannabis	15
Amphetamines	20

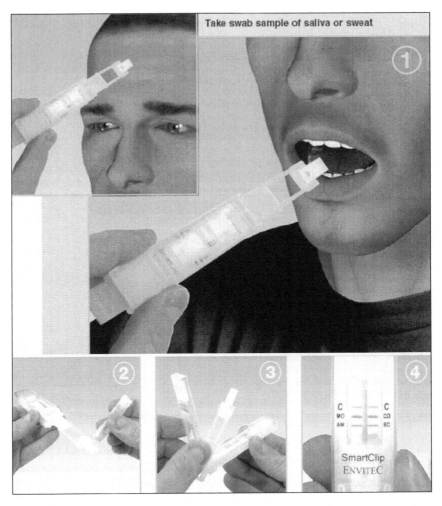

Fig. 21 Sequence of steps for use se of the SmartClip: (1) taking specimen (sweat or saliva), (2) insertion of buffer solution, (3) closing of SmartClip and (4) obtain reading

OraLab® Test for Saliva Testing

In addition, Varian Inc. offers an OraLab® test set for oral fluid measurements, with a patent-pending expresser in a single tube design, which allows the user to collect and store sufficient saliva (can be measured and observed in the tube) for both immediate and confirmation testing (Fig. 22).

OraTube® for Fluid Testing

Another option is the OraTube® for oral fluid testing (Varian company, USA) where the collector is placed inside the donors mouth for 3 min, which saturates

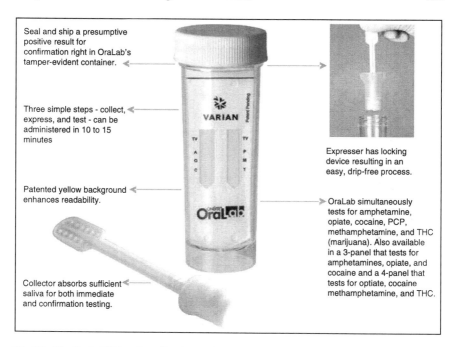

Fig. 22 The OraLab™ Test for collection and storage of saliva for confirmation purposes

and then is placed into the expresser being tightly sealed and then shipped to the laboratory for analysis (Fig. 23).

DRUGWIPE® 5+ for Drug Testing of Saliva

By using an integrated liquid vial, the DrugWipe® 5+ (Securetec Detections-Systems, Munich/Germany) is another option for testing of saliva, being easy to handle and reliable in its use (Fig. 24). DrugWipe® 5+ can detect multiple substances within one sample. Tests have shown that the results of Securetec's rapid saliva-based analysis method are highly consistent (>90%) with the outcome of subsequent blood analysis.

The company has put particular effort into significantly improving the ability of DrugWipe® 5+ to detect THC (cannabis). DrugWipe® 5+ exceeds other roadside test methods, e.g. urine or single-substance tests, in terms of performance and reliability. This was underlined by a broad survey conducted in Germany and other countries where DrugWipe® 5+ was tested in practice, especially its usability in traffic controls and other everyday procedures in police work. The fivefold drug test with an integrated ampoule allows the detection of opiates, cocaine, amphetamines, methamphetamines (e.g. ecstasy) and cannabis (Fig. 25).

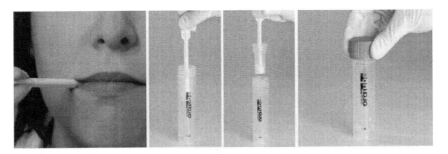

Fig. 23 Application of the OraTube® for oral fluid testing

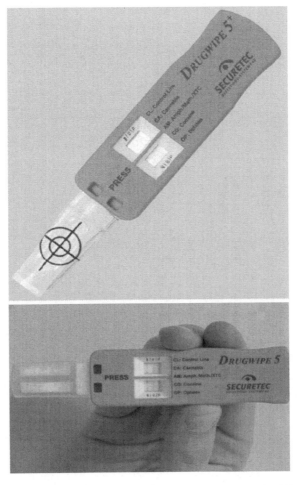

Fig. 24 DrugWipe® 5+, the enhanced version of the proven saliva test DrugWipe® 5

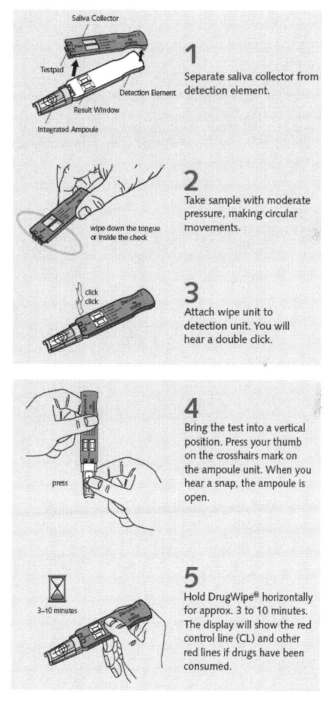

Fig. 25 Sequence of steps for rapid testing of saliva with an integrated liquid ampoule to identify traces of drugs

To support drug enforcement agencies operating at night, a reader for DrugWipe® was developed. The equipment contains an opto-electronic photometer for the analysis of the DrugWipe® test result. The reader measures the color intensity of the displayed lines using a CMOS camera. Depending on the intensity, a value is shown as positive/negative. Using an internal table, all test results either as positive or negative are being displayed. A standard PC (or laptop) can be used as an input and output device for data storage/retrieval.

In addition, single Drug ID test strips (Securetec) are available for the detection of either cannabis, amphetamines, methamphetamines, ecstasy, opiates, or cocaine in unknown powder, liquid or tablets.

Dräger Drug Test 5000

Only recently in 2008 Dräger Company/Germany has introduced the DrugTest 5000 Analyzer, which detects traces of opiates, cocaine, cannabinoids and amphetamines, as well as designer drugs and sedatives from the group of benzodiazepines. The results for five of the drug classes are available after just 5 min, while the test for cannabinoids is completed after 10 min. This not only makes it possible to exclude drug abuse quickly, it also helps to reduce the number of cost- and time-intensive laboratory blood tests. The system comprises two main components, i.e. the Dräger DrugTest 5000 Analyzer (Fig. 26) and the Dräger DrugTest 5000 Test Kit (Fig. 26). After the protective cap has been removed from the saliva test collector, a sample is taken from the patient's mouth. As soon as sufficient saliva has been gathered in the test kit for analysis, the built-in indicator turns blue. The operator then places the test cassette into the analyzer, which displays whether the result is "positive" or "negative" for every drug class on its color display. Acoustic signals support and inform the operator during the entire procedure.

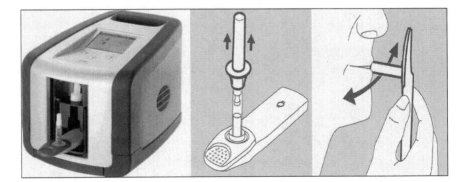

Fig. 26 Dräger Drug Test Analyzer (*left*) and the Dräger Drug Test Kit with protector cap (*right*) for sampling of saliva, with an indicator for sufficient sampling of the specimen

Due to its simple operation, the Dräger DrugTest 5000 is far more discreet and hygienic for both patients and operators than urine-based sample collection. It also minimizes health risks associated with handling body fluids. In addition, controlling the entire sample collection procedure is much easier, virtually eliminating the possibility of manipulation. The analyzer saves the number, course and results of up to 500 tests. It also documents operator and instrument errors. If immediate documentation of protocols is required, the Dräger DrugTest 5000 can be linked to a portable printer using an infrared interface.

Trace Wipe Test for Identification of Cocaine Smuggling or Abuse

In cooperation with Police and Customs Authorities for the determination of cocaine residues, a wipe test, the *"Cocaine Trace Wipe"* (MMC International BV/ The Netherlands) has been developed. It can be used on all surfaces such as inside cars, baggage, containers, parcels, textile such as jeans, etc. (Fig. 27). In case of traces of cocaine, there will be a reaction on the tissue turning into a blue color.

Although the specific determination of an antibody only reacts to an antigen, and similar as in urine drug testing, it can occur that a molecule with a very similar structure causes a reaction, leading to false positive results. Affected persons know this effect and consistently claim that a new medication must be the reason for a positive result. In the case of a positive result, an approximate quantification must be done. Again, as already outlined above, GC/MS (gas chromatography/mass spectroscopy) confirmation analysis helps in the situation, and results should be available within 2 working days after receipt of the sample. Additionally, the detected illicit substance can be further differentiated, for instance for the type of a benzodiazepine and in cases of polydrug use, where LSD and buprenorphine can also be detected.

Professional Equipment for Drug Analysis of Liquids/ Powder/Tablet

For rapid on-site determination of cocaine or its metabolite in urine or other liquid samples a cocaine drug cassette is available from MMC/The Netherlands (Fig. 28).

For a reliable examination of suspicious powder or tablet a single process proofing set is available, where the sample is diluted in tap water and tested with a single step test strip. The suspicious substance or liquid is sampled using a special single-use stick. The residual substance is then diluted with water. Immersing the Drug ID test strip in the solution will detect the presence of the drug. Separate Drug ID strips can be immersed in the same solution to identify additional drugs.

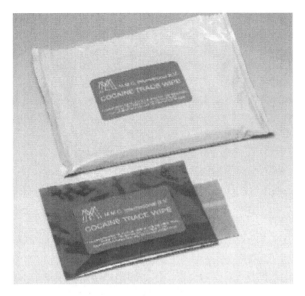

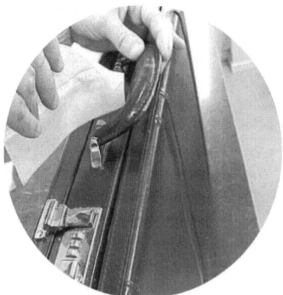

Fig. 27 The Trace Wipe Test for identification of cocaine smuggling or abuse

For professional drug screening a small-sized carrying case is available containing all necessary ingredients for testing of different agents, either being a liquid or a tablet (MMC International/The Netherlands, Fig. 29).

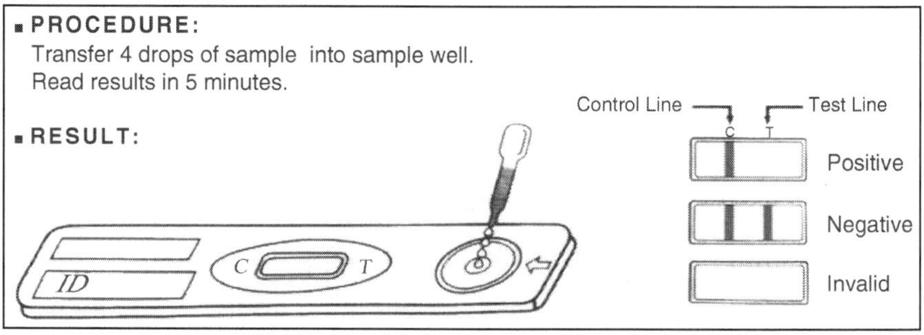

■ **PROCEDURE:**
Transfer 4 drops of sample into sample well.
Read results in 5 minutes.

■ **RESULT:**

Fig. 28 Procedure for on-site rapid determination of cocaine use in liquids

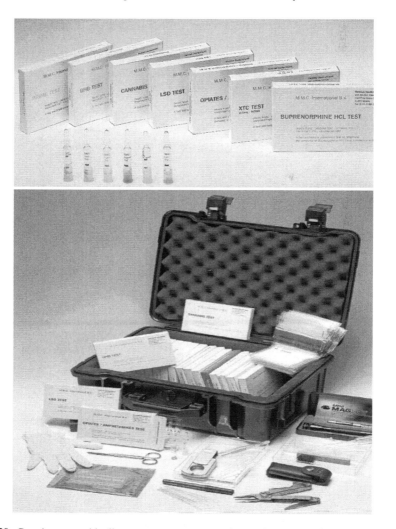

Fig. 29 Carrying vase with all necessary reagents to analyze unknown samples for illicit drugs

Drug Testing in Vapor

Another test system option for the detection of drug vapors in closed shipping containers and other containers is DrugCon® (Securetec Detections-Systems, Munich/Germany; Fig. 30). Sampling probes are available for all container types. With a test time between 5 and 15 min, the analysis of the vapor sample is achieved with a special test cassette in 3–5 min. Because a biochemical detection techniques is used, the failure rate is very low. The core of the container test equipment is an immunochemical cassette with a test strip. The test sample is filtered through this test strip. After the sample has been taken, the test cassette is immersed in water and the result is displayed. The threshold for detection is in the sub-picogram level.

Fig. 30 Use of the DrugCon® device in trucks and containers, which gives the following reading: Negative result, only one line is visible on the display; positive result, two lines are visible on the display

References

1. Schnoll SH, Weaver MF. Addiction and pain. Am J Addict. 2003:S27–S35.
2. National Center on Addiction and Substance Abuse Missed Opportunity: Columbia University, New York CASA National Survey of Primary Care Physicians and Patients on Substance Abuse; 2000.
3. Cherny NI. Opioid analgesics. Comparative features and prescribing guidelines. Drugs. 1996;51:714–737.
4. Cami J, Farré M. Drug addiction - mechanisms of disease. N Engl J Med. 2003;349:975–86.
5. McLellan AT, et al. Drug dependence, a chronic medical illness. Implications for treatment, insurance, and outcomes evaluation. JAMA. 2000;284:1689–1695.
6. Kendler KS, et al. Illicit psychoactive substance use, heavy use, abuse, and dependence in a US population-based sample of male twins. Arch Gen Psychiat. 2000;57:261–269.
7. Jacob T, et al. Genetic and environmental effects on offspring alcoholism. New insights using an offspring-of-twins design. Arch Gen Psychiat. 2003;60:1265–1272.
8. Kreek MJ, Nielson DA, LaForge KS. Genes associated with addiction: alcoholism, opiate, and cocaine addiction. Neuromol Med. 2004;5:85–108.
9. Koob GF, et al. Neurobiological mechanisms in the transition from drug use to drug dependence. Neurosci Biobehav Rev. 2004;27:739–749.
10. Simpson D, et al. Screening for drugs of abuse (II): cannabinoids, lysergic acid diethylamide, buprenorphine, methadone, barbiturates, benzodiazepines and other drugs. Ann Clin Biochem. 1997;34:460–510.
11. Wolff K, et al. A review of biological indicators of illicit drug use, practical considerations and clinical usefulness. Addiction. 1999;94:1279–1298.
12. Hardman JG, Limbird LE, Molinoff PB, Ruddon RW, Gilman AG. In: Goodman & Gilman's the Pharmacological Basis of Therapeutics. 9th edition New York: McGraw-Hill; 1996.
13. Girault JA, Greengard P. The neurobiology of dopamine signaling. Arch Neurol. 2004;61: 641–644.
14. Freye E, Levy JV. Opioids in Medicine - A Comprehensive Review on the Mode of Action and the Use of Analgesics in Different Clinical Pain States. Dordrecht, The Netherlands: Springer Science + Business Media BV; 2008.
15. Portenoy RK. Opioid therapy for chronic nonmalignant pain: a review of the critical issues. J Pain Symptom Manag. 1996;11:203–217.
16. Weaver MF, Jarvis MAE, Schnoll SH. Role of the primary care physician in problems of substance abuse. Arch Intern Med. 1999;159:913–924.
17. Miotto K, et al. Diagnosing addictive disease in chronic pain patients. Psychosomatics. 1996;37:223–235.
18. Savage SR. Assessment for addiction in pain-treatment settings. Clin J Pain. 2002;18:S28–S38.

19. Compton P, Darakjian K, Miotto K. Screening for addiction in patients with chronic pain and "problematic" substance use: evaluation of a pilot assessment tool. J Pain Symptom Manag. 1996;16:355–363.
20. Smith KE, Fenske NA. Cutaneous manifestations of alcohol abuse. J Am Acad Dermatol. 2000;43:1–16.
21. Fiellin DA, Reid MC, O'Connor PG. Outpatient management of patients with alcohol problems. Ann Intern Med. 2000;133:85–827.
22. Skinner HA, et al. Identification of alcohol abuse using laboratory tests and a history of trauma. Ann Intern Med. 1984;101:847–851.
23. Ewing J. Detecting alcoholism: the CAGE questionnaire. JAMA. 1984;252:1905–1907.
24. Minovitz O, Driol M. The sexual abuse, eating disorder and addiction (SEA) triad: syndrome or coincidence ? Med Law. 1989;8:59–61.
25. Vastag B. What's the connection? No easy answers for people with eating disorders and drug abuse. JAMA. 2001;285:1006–1007.
26. Bastiaens L, Francis G, Lewis K. The RAFFT as a screening tool for adolescent substance use disorders. Am J Addict. 2000;9:10–16.
27. Knight JR, et al. Validity of the CRAFFT substance abuse screening test among adolescent clinic patients. Arch Pediatr Adolesc Med. 2002;156:607–614.
28. Sillanaukee P, Olsson U. Improved diagnostic classification of alcohol abusers by combining carbohydrate-deficient transferrin and gamma-glutamyltransferase. Clin Chem. 2001;47:681–685.
29. Papoz L, et al. Alcohol consumption in a healthy population. Relationship to gamma-glutamyl transferase activity and mean corpuscular volume. JAMA. 1981;245:1748–1751.
30. Freye E, Levy JV. Detection of illicit use of opioids in primary care. In: Freye E, Levy JV, editors. Opioids in Medicine – A Comprehensive Review on the Mode of Action and the Use of Analgesics in Different Clinical Pain States. Dordrecht, The Netherlands: Springer Science + Business Media BV; 2008. pp. 412–465.
31. Hammett-Stabler C, Pesce AJ, Cannon DJ. Urine drug screening in the medical setting. Clinica Chimica Acta. 2002;315:125–135.
32. Galloway JH, Marsh ID. Detection of drug misuse - an addictive challenge. J Clin Pathol. 1999;52:713–718.
33. Shults TS. Medical Review Officer's Handbook. North Carolina: Research Triangle Park; 2002.
34. Gourlay DL. Urine Drug Testing in Clinical Practice: Dispelling the Myths & Designing Strategies: Monograph; 2004.
35. Vandevenne M, Vandenbussche H, Verstraete A. Detection time of drugs of abuse in urine. Acta Clinica Belgica. 2000;55:323–333.
36. Casavant MJ. Urine drug screening in adolescents. Pediatr Clin N Am. 2002;49:317–327.
37. Hattab EM, et al. Modification of screening immunoassays to detect sub-threshold concentrations of cocaine, cannabinoids, and opiates in urine: use for detecting maternal and neonatal drug exposure. Ann Clin Lab Sci. 2000;30:85–91.
38. Braithwaite RA, et al. Screening for drugs of abuse. I: opiates, amphetamines and cocaine. Ann Clin Biochem. 1995;32:123–153.
39. Cook JD, et al. The characterization of human urine for specimen validity determination in workplace drug testing: a review. J Anal Toxicol. 2000;24:579–588.
40. Perrone J, et al. Drug screening versus history in detection of substance use in ED psychiatric patients. Am J Emerg Med. 2001;19:49–51.
41. Oyler JM, et al. Identification of hydrocodone in human urine following controlled codeine administration. J Anal Toxicol. 2000;24:530–535.
42. Cunningham JK, Liu LM. Impacts of Federal ephedrine and pseudoephedrine regulations on methamphetamine-related hospital admissions. Addiction. 2003;98:1220–1237.
43. Wyeth-Ayerst. PROTONIX® (pantoprazile sodium) package insert 2004.
44. Roxane Laboratories. Marinol®; Product Information. In: Arky R, editor. Physicians' Desk Reference (PDR). Montvale, NJ: Medical Economics; 1989. pp. 2544–2546.
45. Bosy TZ, Cole BA. Consumption and quantitation of 9-tetrahydrocannabinol in commercially available hemp seed oil products. J Anal Toxicol. 2000;24:562–566.

46. Leson G, et al. Evaluating the impact of hemp food consumption on workplace drug tests. J Anal Toxicol. 2001;25:691–698.
47. Baden LR, et al. Quinolones and false-positive urine screening for opiates by immunoassay technology. JAMA. 2001;286:3115–3119.
48. Zacher JL, Givone DM. False-positive urine opiate screening associated with fluoroquinolone use. Ann Pharmacother. 2004;38:1525–1528.
49. Caplan YH, Goldberger BA. Alternative specimens for workplace drug testing. J Anal Toxicol. 2001;25:396–399.
50. Kintz P, Samyn N. Use of alternative specimens: drugs of abuse in saliva and doping agents in hair. Ther Drug Monit. 2002;24:239–246.
51. Yacoubian GSJ, Wish ED, Pérez DM. A comparison of saliva testing to urinalysis in an arrestee population. J Psychoactive Drugs. 2001;33:289–294.

Appendix
Common Street Names of Cocaine and Methamphetamine used as a Jargon

Jargon is a terminology that relates to a specific activity, profession or a group. Much like slang it develops as a kind of shorthand, to quickly express ideas that are frequently discussed between members of a group. At the same time it is also meant as a barrier to communication for those outside the group. On the other hand in some cases jargon is used as a shibboleth to distinguish between those who belong to a group and those who do not, sometimes called "guild" or "insider" jargon. Those unfamiliar with a subject can often be tagged by the incorrect use of a jargon.

All-American drug	Cocaine
Andean marching powder	Cocaine. The title refers to the area of South America where coca is grown
Angie	Cocaine
Aspirin	Powder cocaine
Aunt	Powder cocaine
Aunt Nora	Cocaine
Balling	Vaginally implanted cocaine
Banano	Marijuana or tobacco cigarettes laced with cocaine
Barbs	Cocaine
Basa	Crack cocaine
Base	Cocaine; crack
Base crazies	Searching on hands and knees for cocaine or crack
Base head	A person who bases
Based out	To have lost control over basing
Basing	Crack cocaine
Basuco (Spanish)	Cocaine; coca paste residue sprinkled on regular or marijuna cigarette
Batman	Cocaine; heroin
Bazooka	Cocaine; combination of crack and marijuana; crack and tobacco combined in a joint; coca paste and marijuana
Bazulco	Cocaine
Beam	Cocaine

(continued)

Appendix (continued)

Term	Definition
Beam me up scottie	Cocaine (powder or crack) combined with PCP
Behind the scale	To weigh and sell cocaine
Beiging	Chemicals altering cocaine to make it appear a higher purity; chemically altering cocaine to make it look brown
Belushi	Combination of cocaine and heroin
Bernice	Cocaine
Bernie	Cocaine
Bernie's flakes	Cocaine
Bernie's gold dust	Cocaine
Big bloke	Cocaine
Big C	Cocaine
Big flake	Cocaine
Big rush	Cocaine
Billie hoke	Cocaine
Biker coffee	Methamphetamine, or crystal meth. It is quite popular in the biker community
Bipping	Snorting heroin and cocaine, either separately or together
Birdie powder	Cocaine; heroin
Black rock	Crack cocaine
Blanca (Spanish)	Cocaine
Blanco (Spanish)	Heroin plus cocaine
Blast	Cocaine; marijuana; smoke marijuana or crack
Blizzard	A white cloud in a pipe being used to smoke cocaine
Blotter	Crack cocaine; LSD
Blow	Cocaine; to inhale cocaine; to smoke marijuana; to inject heroin
Blow blue	To inhale cocaine
Blow coke	To inhale cocaine
Blow smoke	To inhale cocaine
Blunt	Marijuana inside a cigar; cocaine and marijuana inside a cigar
Body-packer	Individual who ingests wrapped packets of crack or cocaine for transport
Booger sugar	Cocaine
Bolivian marching	Cocaine powder
Booster	To inhale cocaine
Bopper	Crack cocaine
Bouncing powder	Cocaine
Boy cocaine	Heroin
Boy-girl	Heroin mixed with cocaine
Break night	Staying up all night on a cocaine binge until daybreak
Brick	Crack cocaine; cocaine; marijuana; 1 kg of marijuana
Bubble gum	Cocaine; crack cocaine; marijuana from Tennessee
Bump crack	Fake crack; cocaine; boost a high; hit of ketamine
Bump up	Use of cocaine to bolster MDMA
Bumper	Crack cocaine
Bumping up	Methylenedioxymethamphetamine (MDMA) combined with powder cocaine
Bunk	Fake cocaine; crack cocaine
Burese	Cocaine
Burnese	Cocaine
Bush	Marijuana; cocaine; PCP
C	Cocaine
C & M	Cocaine and morphine

(continued)

Appendix (continued)

C joint	Place where cocaine is sold
C-dust	Cocaine
C-game	Cocaine
Cabello (Spanish)	Cocaine
Cadillac	Cocaine; PCP
Caine	Cocaine; crack cocaine
California cornflakes	Cocaine
Came	Cocaine
Candy	Cocaine; crack cocaine; amphetamine; depressants
Candy C	Cocaine
Candy flipping on a string	Combining or sequencing LSD with MDMA; mixing LSD, MDMA, and cocaine
Candy sticks	Marijuana cigarettes laced with powdered cocaine
Candy sugar	Powder cocaine
Caps	Heroin; psilocybin/psilocin; crack; γ-hydroxybutyrate
Carnie	Cocaine
Carrie	Cocaine
Carrie nation	Cocaine
Caviar	Combination of cocaine and marijuana; crack cocaine
Cat killer	Presumably a reference to curiosity about the drug, another name for ketamine
CDs	Crack cocaine
Cecil	Cocaine
Chalked up	Under the influence of cocaine
Chalking	Chemically altering the color of cocaine so it looks white
Champagne	Combination of cocaine and marijuana
Charlie	Cocaine
Chase	To smoke cocaine; to smoke marijuana
Chicken scratch	Searching on hands and knees for crack or cocaine
Chippy	Cocaine
Choe	Cocaine
Cholly	Cocaine
Cigamos	Combination of crack cocaine and tobacco
Club drug	A drug that is used by the relatively upscale, generally at nightclubs. Can include ecstasy, ketamine, cocaine and many others drugs
Coca	Cocaine
Cocaine blues	Depression after extended cocaine use
Cocktail	Combination of crack and marijuana; cigarette laced with cocaine or crack; partially smoked marijuana cigarette inserted in regular cigarette; to smoke cocaine in a cigarette
Cocoa puff	To smoke cocaine and marijuana
Coconut	Cocaine
Coke	Cocaine; crack cocaine
Coke bar	A bar where cocaine is openly used
Cola	Cocaine
Combol	Cocaine
Comeback	Benzocaine and mannitol used to adulterate cocaine for conversion to crack
Connie	Powder cocaine
Cooking up	To process powdered cocaine into crack
Coolie	Cigarette laced with cocaine
Copping zone	A place where illicit drugs can be bought. An open air drug market

(continued)

Appendix (continued)

Cork the air	To inhale cocaine
Corrine	Cocaine
Corrinne	Cocaine
Cotton brothers	Cocaine, heroin and morphine
Crack	Cocaine
Crack bash	Combination of crack cocaine and marijuana
Crack gallery	A place where illicit drugs (usually crack) are bought and sold, synonymous with open air drug market
Crisscrossing	The practice of setting up a line of cocaine next to a line of heroin. The user places a straw in each nostril and snorts about half of each line. Then the straws are crossed and the remaining lines are snorted
Croak	Cocaine mixed with methamphetamine; methamphetamine
Crystal	Cocaine; amphetamine; methamphetamine; PCP
Dama blanca (Spanish)	Cocaine
Devil's dandruff	Crack cocaine; powder cocaine
Dirties	Marijuana cigarettes with powder cocaine added to them
Do a line	To snort cocaine
Double breasted dealing	Dealing with cocaine and heroin together
Double bubble	Cocaine
Draf	Marijuana; ecstasy, with cocaine
Dream	Cocaine
Duct	Cocaine
Dust	Marijuana mixed with various chemicals; cocaine; heroin; PCP
Dynamite	Cocaine mixed with heroin
El diablito (Spanish)	Cocaine, marijuana, heroin and PCP
El perico ("parrot")	Cocaine
Electric kool-aid	Crack cocaine
Esnortiar (Spanish)	Cocaine
Everclear	Cocaine; γ-hydroxybutyrate (GHB)
Fast white lady	Powder cocaine
Five-way	Combines snorting of heroin, cocaine, methamphetamine, grind up flunitrazepam pills, and drinking alcohol
Flake	Cocaine
Flame cooking	Smoking cocaine base by putting the pipe over a stove flame
Flamethrowers	Cigarette laced with cocaine and heroin; heroin, or cocaine and tobacco
Flash	LSD; the rush after cocaine injection
Flave	Powder cocaine
Fleece	Counterfeit crack cocaine
Flex	Fake crack (rock cocaine)
Florida snow	Cocaine
Foo foo	Cocaine
Foo foo stuff	Heroin; cocaine
Foo-foo dust	Cocaine
Foolish powder	Cocaine; heroin
Freebase	Rerert to smokable cocaine; crack cocaine
Freeze	Cocaine; to back on a drug deal
Frisco special	Cocaine, heroin, and LSD
Frisco speedball	Composition of cocaine, heroin, and a dash of LSD
Friskie powder	Cocaine

(continued)

Appendix (continued)

Frontloading	The process of transferring a drug solution from one syringe to another
G-rock	1 g rock cocaine
Gaffel	Fake cocaine
Geek-joints	Cigarettes or cigars filled with tobacco and crack; a marijuana cigarette laced with crack or powdered cocaine
Geeze	To snort cocaine
Ghostbusting	Smoking cocaine; searching for white particles in the belief that they are crack
Gift-of-the-sun	Cocaine
Gift-of-the-sun-god	Cocaine
Gin	Cocaine
Girl	Cocaine; crack cocaine; heroin
Girlfriend	Cocaine
Glad stuff	Cocaine
Go on a sleigh ride	To snort cocaine
Gold dust	Cocaine
Goofball	Cocaine mixed with heroin, or anti depressants
Greek	Combination of marijuana and powder cocaine
Greengold	Cocaine
Gremmies	Combination of cocaine and marijuana
H & C	Heroin and cocaine
Half piece	0.5 oz of heroin or cocaine
Handlebars	Combination of crack cocaine and alprazolam
Happy dust	Cocaine
Happy powder	Cocaine
Happy trails	Cocaine
Have a dust	Cocaine
Haven dust	Cocaine
He-she	Heroin mixed with cocaine
Heaven	Cocaine; heroin
Heaven dust	Cocaine; heroin
Henry VIII	Cocaine
Her	Cocaine
Hitch up the reindeer	To snort cocaine
Hooter	Cocaine; marijuana
Horn	To inhale cocaine; crack pipe; to inhale a drug
Horning	To inhale cocaine; heroin
Hunter	Cocaine
Ice	Cocaine; crack cocaine; smokable methamphetamine; methamphetamine; methylenedioxymethamphetamine (MDMA); phencyclidine (PCP)
Icing	Cocaine
Inca message	Cocaine
Jam	Cocaine; amphetamine
Jejo	Cocaine
Jelly	Cocaine
Jim jones	Marijuana laced with cocaine and PCP
Joy powder	Cocaine; heroin
Junk	Cocaine; heroin
King	Cocaine
King's habit	Cocaine
Lace	Cocaine and marijuana

(continued)

Appendix (continued)

Lady	Cocaine; heroin
Lady caine	Cocaine
Lady snow	Cocaine
La dama blanca	Cocaine
Late night	Cocaine
Leaf	Cocaine; marijuana
Line	Cocaine
Liquid lady	Cocaine that is dissolved in water and ingested as a nasal spray or a drink of alcohol + cocaine
Love affair	Cocaine
Ma'a (samoan)	Crack cocaine
Macaroni and cheese	$5 pack of marijuana and a dime bag of cocaine
Mama coca	Cocaine
Marching dust	Cocaine
Marching powder	Cocaine
Mayo	Cocaine; heroin
Merck	Cocaine
Merk	Cocaine
Mighty white	A form of crack cocaine that is hard, white, and pure
Mix	A term used to refer to cocaine or a drug environment
Mo	Marijuana; powder cocaine
Mojo	Cocaine; heroin
Monkey	Cigarette made from cocaine paste and tobacco; drug dependency; heroin
Monos (Spanish)	Cigarette made from cocaine paste and tobacco
Monster	Cocaine
Mosquitos	Cocaine
Movie star drug	Cocaine
Mujer (Spanish)	Cocaine
Murder one	Heroin and cocaine
Nieve (Spanish)	Cocaine
Nose	Cocaine; heroin
Nose candy	Cocaine
Nose powder	Cocaine
Nose stuff	Cocaine
Number 3	Cocaine; heroin
One and one	To snort cocaine
One bomb	100 rocks of crack cocaine
One on one house	Where cocaine and heroin can be purchased
One plus one sales	Selling cocaine and heroin together
Onion	1 oz of crack cocaine
Oyster stew	Cocaine
P-dogs	Combination of cocaine and marijuana
Paradise	Cocaine
Paradise white	Cocaine
Pariba	Powder cocaine
Pearl (flake)	Cocaine
Percia	Cocaine
Perico (Spanish)	Cocaine
Peruvian	Cocaine
Peruvian flake	Cocaine

(continued)

Appendix (continued)

Peruvian marching power	Cocaine
Peruvian lady	Cocaine
Picking	Searching on hands and knees for cocaine or crack
Piece	Cocaine; crack cocaine; 1 oz
Pimp	Cocaine
Polvo blanco (Spanish)	Cocaine
Pop to inhale cocaine	Crack cocaine
Powder	Cocaine; heroin; amphetamine
Powder diamonds	Cocaine
Premos	Marijuana joints laced with crack cocaine
Press	Cocaine; crack cocaine
Primo	Crack; marijuana mixed with cocaine; crack and heroin; mixture of heroin, cocaine and tobacco
Primo turbo	Combination of crack cocaine and marijuana
Primos	Cigarette laced with cocaine and heroin
Pseudocaine	Crack cocaine cut with phenylpropanolamine
Purple caps	Crack cocaine
Purple haze	LSD; crack cocaine; marijuana
Quill	Cocaine; heroin; methamphetamine
Racehorse charlie	Cocaine; heroin
Rane	Cocaine; heroin
Raw	Crack cocaine; high purity heroin
Ready rock	Cocaine; crack cocaine; heroin
Real tops	Crack cocaine
Recompress	Change the shape of cocaine flakes to resemble "rock"
Rider	5 kg of heroin sometimes provided at no cost per 100 kg of cocaine imported from Colombia
Rock star	Female who trades sex for crack or for money to buy crack; a person who uses rock cocaine
Rock(s)	Cocaine; crack cocaine
Roxanne	Cocaine; crack
Rush	Cocaine; isobutyl nitrite; inhalants
Sandwich	Two layers of cocaine with a layer of heroin in the middle
Schmeck	Cocaine
Schoolboy	Cocaine; codeine
Scorpion	Cocaine
Scottie	Cocaine
Scotty	Cocaine; crack; the high from crack
Scramble	Crack cocaine; low purity, adulterated heroin plus crack cocaine
Seconds	Second inhalation of crack from a pipe
Serial speedballing	Sequencing cocaine, cough syrup, and heroin over a 1–2 day period
Serpico 21	Cocaine
Sevenup	Cocaine; crack
Shabu	Combination of powder cocaine and methamphetamine; crack cocaine; methamphetamine; methylenedioxymethamphetamine
Shake	Marijuana; powder cocaine
Shaker/baker/water	Materials needed to freebase cocaine: shaker bottle, baking soda, water
She	Cocaine
Shebanging	Mixing cocaine with water and squirting it up the nose
Sherman stick	Crack cocaine combined with marijuana in a blunt
Shot	To inject a drug; an amount of cocaine; 10 shot or 20 shot
Shrile	Powder cocaine

(continued)

Appendix (continued)

Skin popping	Subcutaneous injection of drugs
Slamming	Intravenous injection of drugs designed to be taken orally
Sleigh ride	Cocaine
Smoke houses	Crack houses
Smoking gun	Pipe for smoking heroin and cocaine
Sniff	To snort cocaine; methcathinone; inhalants
Snort	To inhale cocaine nasally; powder cocaine; use as an inhalant
Snow	Cocaine; heroin; amphetamine
Snow bird	Cocaine user; cocaine
Snow seals	Cocaine and amphetamine
Snow white	Cocaine
Snowball	Cocaine and heroin
Snow cones	Cocaine
Soap	Gamma hydroxybutyrate (GHB); crack cocaine; methamphetamine
Society high	Cocaine
Soda	Injectable cocaine
Soft	Powder cocaine
Soup	Crack cocaine
Space	Crack cocaine
Spaceball	PCP used with crack or powder cocaine
Speedball	Cocaine mixed with heroin; crack and heroin smoked together; methylphenidate (ritalin) mixed with heroin and/or amphetamine
Speedballing	To shoot up or smoke a mixture of cocaine and heroin; ecstasy mixed with ketamine; the simultaneous use of a stimulant with a depressant
Speedballs-nose-style	The practice of snorting cocaine
Splitting	Rolling marijuana and cocaine into a single joint
Sporting	To snort cocaine
Squirrel	Combination of PCP and marijuana, sprinkled with cocaine and smoked; marijuana, PCP, and crack combined and smoked; LSD
Stacks	Methylenedioxymethamphetamine adulterated with heroin or crack
Star	Cocaine; amphetamine; methcathinone
Star-spangled powder	Cocaine
Stardust	Cocaine; PCP
Studio fuel	Cocaine
Sugar	Cocaine; crack cocaine; heroin; LSD
Sugar boogers	Powder cocaine
Sweet stuff	Cocaine; heroin
T	Cocaine; marijuana
Talco (Spanish)	Cocaine
Tardust	Cocaine
Teenager	Cocaine
Teeth	Cocaine; crack cocaine
The five way	Heroin plus cocaine plus methamphetamine plus Rohypnol® (flunitrazepam) plus alcohol
Thing	Cocaine; crack cocaine; heroin; the main drug of interest at the moment
Tio	Cocaine-laced marijuana cigarette
Toke	To inhale cocaine; to smoke marijuana; marijuana
Toot	Cocaine; to inhale cocaine
Toot	Perika®, brand name for extract of St John Wort (anti depressant effect) plus cocaine

(continued)

Appendix (continued)

Trails	Cocaine; LSD induced perception that moving objects leave multiple images or trails behind them
Trey	Small rock of crack cocaine
Turkey	Cocaine; amphetamine
Tutti-frutti (Portuguese)	Flavored cocaine
Twinkie	Crack cocaine
Uptown	Powder cocaine
White boy	Heroin; powder cocaine
White dragon	Powder cocaine; heroin
White girl	Cocaine; heroin
White horse	Cocaine; heroin
White lady	Cocaine; heroin
White mosquito	Cocaine
White powder	Cocaine; PCP
Whiz bang	Cocaine; heroin and cocaine
Wicky	Combination of powder cocaine, PCP and marijuana
Wild cat	Methcathinone mixed with cocaine
Window pane	LSD; crack cocaine
Wings	Cocaine; heroin
Witch	Cocaine; heroin
Woo blunts	Marijuana; marijuana combined with cocaine
Woolas	Cigarettes laced with cocaine; crack sprinkled on marijuana cigarette
Woolie	Marijuana and heroin combination; marijuana and crack cocaine combination
Woolie blunt	Combination of crack cocaine and marijuana
Wooties	Crack smoked in marijuana joints
Working bags	Bags containing several small rocks of crack cocaine
Yam	Crack cocaine
Yao	Powder cocaine
Yay	Rock of crack cocaine
Yeyo	Cocaine, Spanish term
Zip	Cocaine

Index

A
Aberrant drug use, demask, 243ff
Amphetamine, methamphetamine, 109ff
   appetite suppressants, 135
Analysis of saliva, hear and sweat, 269ff
   analysis of sweat, 271
   Dräger drug test 500, 278
   drug testing in vapor, 282
   DrugWipe® 5+ for saliva testing, 275, 276
   OralLab® test for saliva testing, 274
   OraTube® for fluid testing, 274–276
   SmartClip® test for saliva testing, 272–274
   Trace wipe test, 279, 280

C
Chronic toxicity, cocaine, 69
Craving, 80, 81, 85, 88, 89, 93, 94, 96, 99, 101, 104
Coca leaf, 6, 7, 9–12, 13, 15–18, 20, 23, 25, 29–31, 33, 34, 49, 56, 257
   botanical characters, 7
   CoCa de Mate, 12
   production, 9
Coca making, jungle, 29
   drying of coca paste, 29
   pressing, 30
   separation from kerosene, 30
   soaking of coca leaf, 29
Coca plant ingredients, 27
   benzoylecgonine, 27
   cinnamylcocaine, 27
   cocaine, 27
   cocamine, 27
   ecgonine, 27
Cocaethylene, 51, 52, 69
Cocaine, 5ff
   addictive properties, 61
      addiction, 63
      dopaminergic reward system, 62
      physical dependence, 63
      psychological dependence, 64
   depression, 80, 85, 89, 90, 92–94, 103
   diluting (cutting), 36
   paranoia, 79, 80, 83–85
   forms, 25
      coca leaf, 25
      coca paste, 25
      cocaine base, 25
      crack cocaine, 25
      powder cocaine, 25
   history, 13
      coca wine (Vin Mariani), 14
      Coca-Cola, 5, 16–19
      Controlled Substances Act of 1970,
      French wine coca, 15, 17
      Harrison Narcotic Act, 17
      Pemberton, John Styth, 15–17
      Pure Food & Drug Act, 49, 114
      regulatories, 16, 17
   incidence of illicit use, 1
   intoxication treatment, 65
      basics in treatment, 65
      Casey Jones reaction, 65
   medicinal use, 19
      Freud, Sigmund, 18–21, 54, 58, 60
      local anesthetic properties, 21
      Erlanger-Gasser, 23, 58
      Koller, Karl, 21, 22, 51
      mechanism of action, 51
      magistral formularies, 24
      von Fleisch-Marxow, Dr. Ernst, 21, 54
   overdose (OD), 66, 67
      cerebrovascular ischemia/hemorrhage, 74
      delirium, 68
      hyperthermia, 65, 68, 76
      hypertonia, 65
   kindling, 80, 81, 84

297

Cocaine (cont.)
  myocardial infarction, 70, 71
  rhybdomylosis, 73
  seizures, 66, 67, 72, 76, 80, 84
  tachyarrhythmia, 75, 76
  ventricular fibrillation, 67
  pharmacology, 49
    central nervous effects, 54
      mechanism of action, 54
    decomposition, 51
    lethal doses, 77
    molecular structure, 50
    physicochemical properties, 50
    toxicity, 53
  pregnancy, 87–97
Crystal methamphetamine, 125
  addictive properties, 125
  plasma levels, 126, 128
Chronic use of MDMA, 156ff
  neurotoxicity, 156–159
    Ricaurte, George, 157, 158sCutt-offs in urine drug screening, 255

D

Designer drugs, 181ff
  3-methylfentanyl, 187, 188
  acetyl-alpha-methylfentanyl, 187
  alpha-menthylfentanyl, 187
  benzylfentanyl, 187, 188
  history, 183
  MPPP or 1-phenyl-4-methyl-4-hydroxypiperidine, 85, 187
  para-fluorofentanyl, 187, 188
  piperazine-based, 189
  Shulgin, Alexander, 185, 186
  tryptamine-based, 187
Diluting (cutting) of cocaine, 36

E

Ecstasy *see* MDMA
Enzyme linked immune sorbent assay (ELISA), 253
Erythroxylum coca, 5, 6
Erythroxylum Novogranatense, 5, 6, 18, 27

F

Freebasing of cocaine, 43–47
  baking soda method, 47
  ether wash method, 44, 45
Foxy, methoxy (5-MeO-DIPT), 233
  chemical structure, 234

  legal status, 234
  pharmacology, 233

G

Gamma-butyrolactone (GBL), 189, 192, 201, 204, 207
Gamma-hydroxybutyric acid (GHB, liquid ecstasy), 191
  abuse potential and intoxication, 203
    body temperature, 206
    cardiovascular effects, 205
    gastrointestinal effects, 206
    metabolic effects, 206
    muscular disorders, 207
    ocular effects, 206
    psychotic effects, 206
    respiratory effects, 205
    treatment of GHB overdose, 208–210
    withdrawal and tolerance, 207
  alcohol and opiate withdrawal, 200
  cardiovascular effects, 197
  CNS effects, 196, 197
  history, 191
  medical use, 197, 199
  mode of action, 195–197
  molecular formula, 196
  neuroendocrine effects, 198
  respiratory effects, 198
  sleep pattern, 198
Gas chromatography (GC), 208, 252–254, 259, 260, 270, 279

H

Half-life of detection in urine drug screening, 255

I

Ibogaine, 225
  anti-addictive properties, 225, 226
  chemical structure, 226
  pharmacology, 226
Interpretation of urine drug tests (UDT), 256
Intoxication with MDMA, see MDMA

L

Lysergic acid diethylamide (LSD-25), 229
  chemical structure, 230
  Hoffmann, Albert, 230
  legal status, 231

Index

pharmacology, 229
research, 230

M

Manipulation of urine samples, 267, 268
Mass spectrometry (MS), 254
MDMA (ecstasy), 139–174
  abuse liability, 153
  adulterants, 164–166
  analogs of 173
    MDA or 3,4-methylenedioxyamphetmine, 173
    MDE, MDEA or N-ethyl-methylenedioxyamphetamine, 173
    MMDA or 3-methoxy-4,5,-methylenedioxyamphetamine, 173
    2-methoxy-3,4-methylenedioxyamphetamine, 173
    MBDB or N-methyl-1-(1,3-benzodioxol-5-yl)-N-methylbutan, 174
  autoradiography, SPECT analysis, 149, 157
  effects, MDMA, 157ff
    cardiovascular, 154
    empathogenesis, 152
    entactogenesis, 152
  incidence of illicit use, 143
  intoxication, 165
    hepatotoxicity, 168
    psychopathology, 168
    treatment of overdose, 168
      hypertension, 170
      hyperthermia, heatstroke, 169
      intracranial hemorrhage, infarction, 170, 171
      myoclonus, 171
      neuroleptic malignant syndrome, 171
      restlessness, 171
      rhabdomyolysis, 171
      serotonin syndrome, 171
  lethal doses, 153
  medical use in psychotherapy, 143
  molecular structure, 141, 148
  neuroendicrine, 154
  ocular, 155
  Parkinson, 145
  pharmacokinetics of, 161
    primary, 161
    secondary, 163
  pharmacology, 147
  physical effects, 153
  post traumatic stress syndrome (PTSD), 140, 144
  serotonin transporter (SERT), 147, 149

benzylpiperazine (BZP), 211
  addictive effects, 212
  pharmacology, 211
cerebral protection, 199
Datura Stramonium, 215
  overdose, 217
  toxicity, 217
dimethyltryptamine (DMT), 219
  chemical structure, 220
  hallucinogenic effects, 219
narcolepsy and insomnia, 200
pharmacokinetics, 194
  metabolism of GHB, 201
Psilocybin, 221
  chemical structure, 222
  pharmacological effects, 221
  toxicity, 221
sedation in anesthesia, 199
Methadone metabolite (EDDP), 255, 262–265
Methamphetamine (speed), 109ff
  addiction, 131–133
    non-medical treatment, 132
    prevention of relapse, 132
    treatment, 131
  Controlled Substances Act of 1970, 114
  Drug Prevention and Misuse act, 112
  effects, 120
  history, 109
  mode of action,115
  Panzerschokolade, 110
  Pervitin, 110, 111
  pharmacology, 119–124
  physicochemical properties, 124
  prescribing, 122
  STUKA-Pills, 110

O

Opiates, 16, 17, 19, 255, 256, 259–261, 265, 272, 275, 278

P

Pharmacokinetics of MDMA, see MDMA
Phencyclidine (PCP) analogues, 189
Piperazine-based designer drugs, 189

S

Signs and symptoms of drug abuse, 244ff
  alcohol addiction, 244
    alanine transferase (ALT), 246
    CAGE questionnaire, 244
    CRAFFT test, 245

Signs and symptoms of drug abuse (*cont.*)
    gamma glutamate transferase
        (γ GT), 245
    mean corpuscular volume (MCV), 245
    Trauma test, 244
  skin popping, 247, 249, 250
Steps ascertaining purity of cocaine, 37
Steps in refining cocaine, 31, 32
  coca base, 32
  cocaine hydrochloride (crystal), 32
Street names of cocaine and
    methamphetamine, 287

T

Trafficking of cocaine, 33ff
  body packer, 35
  smuggling of cocaine, 34, 35, 279
Treatment options of cocaine abuse, 88
  cocaine free environment, 89
  condition behavioral therapy (CBT), 89
  detoxifixation, 88
  pharmacological treatment, 89
    amantadine, 94
    antidepressants, 90
    Bupropion, 90
    disulfiram, 100
    domapine reuptake inhibitors, 103
    dopamine antagonists, 93
    gamma vinyl-GABA, 99
    L-dopa ligands, 95
    lithium, 96
    L-tryptophan, 90
    L-tyrosine, 90
    methylphenidate, 95
    vaccination, 102
Tryptamine-based designer drugs, 188
Type of test in urine drug testing, 261
  dip test, 261
  drop test, 261
  test cups, 265–267

U

Urine drug screening (UDS), 251
  prerequisites, 251